MATHEMATICS IN INDUSTRY   **28**

More information about this series at http://www.springer.com/series/4650

Ulrich Langer • Wolfgang Amrhein •
Walter Zulehner

Editors

Scientific Computing in Electrical Engineering

SCEE 2016, St. Wolfgang, Austria,
October 2016

 Springer

Editors
Ulrich Langer
Institute of Computational Mathematics
(NuMa)
Johannes Kepler University (JKU)
Linz, Austria

Wolfgang Amrhein
Institute for Electric Drives and Power
Electronics
Johannes Kepler University (JKU)
Linz, Austria

Walter Zulehner
Institute of Computational Mathematics
(NuMa)
Johannes Kepler University (JKU)
Linz, Austria

ISSN 1612-3956   ISSN 2198-3283   (electronic)
Mathematics in Industry
ISBN 978-3-030-09259-7   ISBN 978-3-319-75538-0   (eBook)
https://doi.org/10.1007/978-3-319-75538-0

Mathematics Subject Classification (2010): 65-06, 65Lxx, 65Mxx, 65Nxx, 65L06, 65L12, 65L15, 65L60, 65L80, 65M06, 65M60, 78-06

Printed on acid-free paper

This Springer imprint is published by the registered company Springer International Publishing AG part of Springer Nature.
The registered company address is: Gewerbestrasse 11, 6330 Cham, Switzerland

Foreword

In the name of Johannes Kepler University Linz I have the honor to welcome you to the 11th international conference on scientific computing in electrical engineering.

I have learned your conference is a very special one. You are not only connecting people, you are at the same time connecting scientific disciplines and building bridges between science and industry. Almost two decades of successfully bringing together applied mathematicians and electrical engineers prove the initial idea of the conference right. What started with visionary and very ambitious goals is now fully realized and working out well. I congratulate your community on your pioneering spirit, on your collaborative approach, and of course on consistently pursuing your goals over the years.

I recognized a few similarities to the last five decades of JKU. In this academic year Johannes Kepler University celebrates its 50th anniversary. JKU is a relatively young university but though has an eventful history. Generations of courageous and persistent JKU teachers and researchers have contributed to what the university is today: a vivid place of innovation, education, and know-how transfer.

LCM and RICAM—your conference hosts this year—are two institutions at our university we are proud of.

Mathematicians fostered the establishment of the mechatronics department and supported Europe's first academic degree program in mechatronics, which was created at our university in Linz in 1990. It led to the successful Comet K2 center Linz Center of Mechatronics. Currently there are intense preparations going on for the next funding period of this Comet competence center.

The structure of RICAM serves a little bit as a template for new organizational units we want to develop at JKU. We have just founded the Linz Institute of Technology (LIT) with the aim of bundling our technological competence across all disciplines. The LIT focuses on practically oriented research in all areas of science and technology along the entire innovation chain. We want to foster interdisciplinary collaboration to solve problems of tomorrow and also deal with the societal consequences of constant technological innovation.

The LIT will also be the umbrella brand for all our technological degree programs. Research-guided teaching is of great importance. Students should become involved in research groups and applied project work as early as possible. The RICAM concept of special semesters I think is just great. It sets a pattern for what we want to develop for JKU students on a broad basis.

I have heard that this conference also serves as the kick-off meeting for the next RICAM Special Semester on Computational Methods in Science and Engineering held from October to December. I wish you all the best for your work during the special semester.

For the conference, I wish you inspiring discussions and successful sessions with many new ideas for your further work. I thank the conference chair and the editors of the SCEE proceedings Prof. Ulrich Langer, Prof. Wolfgang Amrhein, and Prof. Walter Zulehner as well as the scientific and the organizational committee for your efforts. Have a good time here at the beautiful Wolfgangsee.

Johannes Kepler University Linz   Meinhard Lukas
Linz, Austria   Rector
October 2016

Preface

The 11th International Conference on "Scientific Computing in Electrical Engineering" (SCEE) was held at the Federal Institute for Adult Education (BIfEB—Bundesinstitut für Erwachsenenbildung) in St. Wolfgang/Strobl, Austria, from October 3 to October 7, 2016. The SCEE 2016 was jointly organized by the doctoral program "Computational Mathematics" and the Institute of Computational Mathematics at the Johannes Kepler University Linz, the Linz Center of Mechatronics GmbH, and the Johann Radon Institute for Computational and Applied Mathematics (RICAM) at the Austrian Academy of Sciences.

With more than 80 scientists from 15 countries participating, the conference brought together applied mathematicians and electrical engineers, academics and industry practitioners, and, last but not least, different communities, namely those working in electromagnetic field computation and those working in circuit and device simulation. The BIfEB creates an inspiring, "Oberwolfach-like" work atmosphere. All talks and the poster introductions were presented in the plenary session in order to avoid splitting up the different communities mentioned above. This led to fruitful discussions both within and across the communities.

The scientific committee invited eight experts to give keynote presentations on the main topics in the regular program. Our keynote speakers were (in alphabetical order):

Ram Achar (Ottawa, Ontario, Canada), Challenges and Opportunities: Modeling and Simulation for the Emerging High-Speed Multifunction Designs,

Peter Benner and Lihong Feng (Magdeburg, Germany), Parametric Model Order Reduction for ET Simulation in Nanoelectronics,

Hans-Georg Brachtendorf (Hagenberg, Austria), Coupled Multirate Simulation by the MPDE Technique for Radio Frequency Circuits,

Carlo de Falco (Milano, Italy), Numerical Modeling of Organic Electronic and Photovoltaic Devices,

Victorita Dolean (Glasgow, UK), Microwave Tomographic Imaging of Cerebrovascular Accidents by Using High-Performance Computing,

Eric Keiter, Albuquerque (New Mexico, USA), Gradient-Enhanced Polynomial Chaos Methods for Circuit Simulation,

Roland Pulch, Greifswald (Germany), Global Sensitivity Analysis for Parameter Variations in Electric Circuits,

Joachim Schöberl (Vienna, Austria), Mapped Tent-Pitching Methods for Maxwell Equations.

Moreover, we organized an industrial day where five scientists from industry presented challenging industrial problems and discussed their solution. These speakers were (in alphabetical order):

Massimiliano Cremonesi (Polimi, Milano, Italy), A Lagrangian Finite Element Approach for the Simulation of a Vacuum Arc,

Lars Kielhorn (TailSiT GmbH, Graz, Austria), A Symmetric FEM-BEM Formulation for Magnetostatics,

Stefan Reitzinger (CST, Darmstadt, Germany), Broadband Solution Methods for Maxwell's Equations in Laplace Domain,

Ehrenfried Seebacher (austriamicrosystems, Unterpremstätten, Austria), Compact Modeling for HV CMOS Technologies,

Siegfried Silber (LCM GmBH, Linz, Austria), Optimization of Mechatronic Components with MagOpt.

In addition to the invited keynote talks, there were also 19 contributed talks. Twenty-six posters were presented and discussed in two poster sessions; the first poster session was devoted to applications including industrial applications, while the second poster session collected posters presenting new computational methods. Each session started with a fast-forward presentation of the posters by one of the authors.

For our excursion on Wednesday we planned to visit the "Five Fingers" on Krippenstein mountain and Hallstatt town and lake. However, weather conditions were so bad that we had to visit the "Giant Ice Cave" near Hallstatt instead. We took the first section of the cable car to Krippenstein, then walked for 20 min to the cave, where we had an hour-long guided tour before returning to the BIfEB—without getting too wet. The picture shows the participants in front of the main lecture hall at the BIfEB (see Fig. 1). More pictures including photos of the excursion can be found on the conference homepage

https://www.ricam.oeaw.ac.at/events/conferences/scee2016/

We would like to thank Nicodemus Banagaaya and Ewald Lindner for providing these pictures.

The contributions to these proceedings are divided into six parts:

I Computational Electromagnetics,
II Circuit and Device Modeling and Simulation,
III Coupled Problems and Multi-Scale Approaches in Space and Time,
IV Mathematical and Computational Methods Including Uncertainty Quantification,

Fig. 1 Participants in front of the main lecture hall at the BIfEB

 V Model Order Reduction,
VI Industrial Applications.

We very much hope that this collection of papers will be of interest to many applied mathematicians and electrical engineers working at universities and research institutions as well as to scientists working in industry. We would like to thank all participants for their valued contributions to the SCEE 2016 and, in particular, we are grateful to the authors of the papers published in these proceedings.

Linz, Austria   Ulrich Langer
Linz, Austria   Wolfgang Amrhein
Linz, Austria   Walter Zulehner
December 2017

Acknowledgements

We would like to thank the Doctoral Program "Computational Mathematics" and the Institute of Computational Mathematics at the Johannes Kepler University Linz, the Linz Center of Mechatronics GmbH, and the Johann Radon Institute for Computational and Applied Mathematics (RICAM) at the Austrian Academy of Sciences for the financial support and their help in the organization of the conference.

We are also grateful for the donation by the GAMM that allowed us to support young scientists. We acknowledge the support given by the company Computer Simulation Technology (CST) for the industrial day, the poster sessions, and the social program. Special thanks go also to the partners of the European project nanoCOPS, Nanoelectronic COupled Problems Solution (FP7-ICT-2013-11/619166, www.fp7-nanocops.eu), for their support of the industrial day.

Last but not least we would like to thank the members of the Local Organizing Committee and the Scientific Committee who helped us very much in preparing and running the conference. The careful reviewing process was only possible with the help of the members of the Scientific Committee who were handling the reviewing process. The anonymous referees did a wonderful work that helped the authors to improve the quality of their contributions. Finally, we express our thanks to Ruth Allewelt from Springer Heidelberg for continuing support and patience while preparing this volume.

Contents

Part I Computational Electromagnetics

1 Preliminary Numerical Study on Electrical Stimulation at Alloplastic Reconstruction Plates of the Mandible 3
Ursula van Rienen, Ulf Zimmermann, Hendrikje Raben, and Peer W. Kämmerer
  1.1   Medical Background and Previous Work 3
  1.2   Set-Up and Solution of the Bio-Electric Model.................... 5
  1.3   Comparison of Two Stimulation Sites............................ 8
  1.4   Conclusion and Outlook 9
  References.. 10

2 Evaluation of Capacitive EMG Sensor Geometries by Simulation and Measurement 13
Theresa Roland, Sabrina Mairhofer, Wolfgang Roland,
Christian Diskus, Sebastian Amsuess, Michael Friedrich Russold,
Christoph Wolf, and Werner Baumgartner
  2.1   Introduction... 14
  2.2   Simulation of EMG 14
      2.2.1   Simulation in COMSOL Multiphysics 14
      2.2.2   Simulation in MATLAB (Nodal Analysis) 15
  2.3   Experimental Methods 17
      2.3.1   Measurement Setup....................................... 17
      2.3.2   EMG Measurement....................................... 18
  2.4   Results 19
  2.5   Conclusion.. 22
  References.. 23

3 Stability Analysis of Electromagnetic Transient Simulations.......... 25
Wim Schoenmaker, Christian Strohm, Kai Bittner, Hans Georg
Brachtendorf, and Caren Tischendorf
  3.1   Introduction.. 25

3.2   Discretization Procedure for the Maxwell-Ampere Equation   26
            3.2.1   Spatial Discretization   27
     3.3   Some Theoretical Considerations   31
     3.4   The Impact of Meshing   32
     3.5   Conclusion   33
     References   33

4   **Sensitivity of Lumped Parameters to Geometry Changes
     in Finite Element Models**   35
     Sebastian Schuhmacher, Carsten Potratz, Andreas Klaedtke,
     and Herbert De Gersem
     4.1   Introduction   35
     4.2   Extraction and Sensitivity Analysis of Circuit Parameters   36
            4.2.1   Partial Inductances   36
            4.2.2   Capacitances and Conductances   38
     4.3   Application Examples   38
            4.3.1   Plate Capacitor   38
            4.3.2   Conducting Wire   39
            4.3.3   Low-Pass π-Filter   40
     4.4   Conclusions   41
     References   42

5   **Electro-Thermal Simulations with Skin-Layers and Contacts**   43
     Christoph Winkelmann, Raffael Casagrande,
     Ralf Hiptmair, Philipp-Thomas Müller, Jörg Ostrowski,
     and Thomas Werder Schläpfer
     5.1   Background   43
     5.2   Electric Contacts   44
     5.3   Adaptive Refinement for Ohmic Losses   46
     5.4   Approximation Quality on Coarse Meshes   47
            5.4.1   Theory   47
            5.4.2   Numerical Experiments   48
     5.5   Electro-Thermal Simulation   50
     5.6   Conclusions   52
     References   52

Part II   Circuit and Device Modeling and Simulation

6   **Gradient-Enhanced Polynomial Chaos Methods for Circuit
     Simulation**   55
     Eric R. Keiter, Laura P. Swiler, and Ian Z. Wilcox
     6.1   Introduction   55
     6.2   Transient Sensitivities   56
            6.2.1   Transient Direct Sensitivities   58
            6.2.2   Transient Adjoint Sensitivities   58
     6.3   Polynomial Chaos Expansion Methods   60

6.4   Results for CMOS Inverter Circuit 63
6.5   Conclusions 66
References 67

7   Coupled Circuit Device Simulation 69
Kai Bittner, Hans Georg Brachtendorf, Wim Schoenmaker,
Christian Strohm, and Caren Tischendorf
7.1   Introduction 69
7.2   Coupled Simulation 70
7.3   Implementation 70
        7.3.1   Lumped Device Models 71
        7.3.2   Replacement of a Lumped Device Model
                 by a Field Model 72
7.4   Numerical Results 73
7.5   Conclusions 75
References 77

**Part III   Coupled Problems and Multi-Scale Approaches in Space
                and Time**

**8   Density Estimation Techniques in Cosimulation Using
       Spectral- and Kernel Methods** 81
Kai Gausling and Andreas Bartel
8.1   Introduction 81
8.2   Lower Bound Estimator for Purely Algebraic Coupling 82
8.3   Density Estimation Techniques 83
8.4   Numerical Test Example 85
8.5   Conclusions 88
References 89

**9   Multirate DAE/ODE-Simulation and Model Order Reduction
       for Coupled Field-Circuit Systems** 91
Christoph Hachtel, Johanna Kerler-Back, Andreas Bartel,
Michael Günther, and Tatjana Stykel
9.1   Introduction 91
9.2   Model Order Reduction for Magneto-Quasistatic Equations 93
9.3   Multirate Time Integration for ODE/DAE-Systems 94
9.4   Simulation of a Coupled Electric Field-Circuit System 96
9.5   Conclusions 99
References 100

**10   Modelling and Simulation of Electrically Controlled Droplet
       Dynamics** 101
Yun Ouédraogo, Erion Gjonaj, Thomas Weiland, Herbert De Gersem,
Christoph Steinhausen, Grazia Lamanna, Bernhard Weigand,
Andreas Preusche, and Andreas Dreizler
10.1   Introduction 101

10.2   Numerical Model...   102
          10.2.1   Fluid Problem..   102
          10.2.2   Interface Capturing   103
          10.2.3   Electric Problem   104
10.3   Application and Results..   105
          10.3.1   Electrically Driven Droplet Detachment   105
          10.3.2   Charged Droplet Detachment   107
10.4   Conclusion..   108
References...   108

**Part IV   Mathematical and Computational Methods Including
             Uncertainty Quantification**

**11   Multirate Shooting Method with Frequency Sweep for Circuit
       Simulation**   113
       Kai Bittner and Hans Georg Brachtendorf
       11.1   Introduction...   113
       11.2   Circuit Equations and Multistep Methods   114
       11.3   Periodic Steady States and Shooting   115
       11.4   Multirate Shooting Method   117
       11.5   Frequency Sweep and Smoothness Conditions   118
                 11.5.1   An Explicit Approach   119
                 11.5.2   An Additional Equation   120
                 11.5.3   A Discrete Smoothness Criterion   121
       11.6   Numerical Test   123
       11.7   Conclusion...   124
       References..   124

**12   A Trefftz Method for the Time-Harmonic Eddy
       Current Equation**...   127
       Raffael Casagrande, Christoph Winkelmann, Ralf Hiptmair,
       and Jörg Ostrowski
       12.1   Introduction...   127
       12.2   Non-symmetric Weighted Interior Penalty Framework   128
       12.3   Enriched Approximation Space   131
       12.4   Numerical Example   132
       12.5   Concluding Remarks   134
       References..   135

**13   Survey on Semi-explicit Time Integration of Eddy
       Current Problems**   137
       Jennifer Dutiné, Markus Clemens, and Sebastian Schöps
       13.1   Introduction...   137
       13.2   Mathematical Formulation.....................................   138
       13.3   Numerical Validation...   141

13.4   Conclusion   144
References   145

14   **A Local Mesh Modification Strategy for Interface Problems
       with Application to Shape and Topology Optimization**   147
      Peter Gangl and Ulrich Langer
      14.1   Motivation   147
      14.2   A Local Mesh Modification Strategy for Interface Problems   148
              14.2.1   Preliminaries   149
              14.2.2   Description of the Method   149
      14.3   Condition Number   152
      14.4   Numerical Results   153
      References   155

15   **Numerical Methods for Derivative-Based Global Sensitivity
       Analysis in High Dimensions**   157
      Qingzhe Liu and Roland Pulch
      15.1   Introduction   157
      15.2   The Stochastic Model   158
              15.2.1   Linear Dynamical Systems   158
              15.2.2   Derivative-Based Measures   159
      15.3   Numerical Approaches   160
      15.4   Application Example   161
      15.5   Conclusions   165
      References   167

16   **Fitting Generalized Gaussian Distributions for Process
       Capability Index**   169
      Theo G. J. Beelen, Jos J. Dohmen, E. Jan W. ter Maten,
      and Bratislav Tasić
      16.1   Introduction   169
      16.2   Numerical Results   173
      16.3   A Quality Measure Index for a Generalized Gaussian
              Distribution   175
      16.4   Conclusions   175
      References   176

17   **Robust Optimization of an RFIC Isolation Problem Under
       Uncertainties**   177
      Piotr Putek, Rick Janssen, Jan Niehof, E. Jan W. ter Maten,
      Roland Pulch, Michael Günther, and Bratislav Tasić
      17.1   Introduction   178
      17.2   Modeling Approach   179
              17.2.1   Field Model of the Integrated Circuit   179
              17.2.2   Equivalent Circuit Model and
                        Floorplanning/Grounding Strategy   180
      17.3   Uncertainty Quantification Analysis   182

17.4   Robust Optimization Problem 183
17.5   Numerical Example and Conclusions 183
References... 185

Part V   Model Order Reduction

**18   Sparse Model Order Reduction for Electro-Thermal Problems
with Many Inputs**.. 189
Nicodemus Banagaaya, Lihong Feng, Wim Schoenmaker, Peter
Meuris, Renaud Gillon, and Peter Benner
18.1   Introduction... 190
18.2   BDSM-ET Method for ET Coupled Problems
with Many Inputs....................................... 191
18.3   Proposed Modified BDSM-ET Method 194
18.4   Numerical Experiments 197
18.5   Conclusion.. 201
References... 201

**19   Quadrature Methods and Model Order Reduction for Sparse
Approximations in Random Linear Dynamical Systems**.............. 203
Roland Pulch
19.1   Introduction... 203
19.2   Stochastic Modelling and Sparse Representations................. 204
19.2.1   Linear Dynamical Systems.................................. 204
19.2.2   Stochastic Modelling and Orthogonal Expansions........ 205
19.2.3   Sparse Approximations................................... 206
19.3   Quadrature Methods....................................... 207
19.3.1   Weakly Coupled Linear Dynamical System.............. 207
19.3.2   Symmetric Probability Distributions 208
19.4   Model Order Reduction 209
19.4.1   Sparse Approximation by Model Order Reduction 209
19.4.2   Numerical Rank Deficiency of Output Matrix 211
19.4.3   Error Estimates 211
19.5   Illustrative Example 212
19.6   Conclusions.. 216
References... 216

**20   POD-Based Reduced-Order Model of an Eddy-Current
Levitation Problem**... 219
Md. Rokibul Hasan, Laurent Montier, Thomas Henneron,
and Ruth V. Sabariego
20.1   Introduction... 219
20.2   Magnetodynamic Levitation Model 221
20.3   POD-Based Model Order Reduction 222
20.3.1   Application to an Electro-Mechanical Problem
with Movement 223

20.4   Application Example 225
    20.4.1   RO Modelling with Automatic Remeshing Full
          Domain 226
    20.4.2   RO Modelling with Mesh Deformation
          of a Sub-domain 227
20.5   Conclusion 228
References 228

**21   Time-Domain Reduced-Order Modelling of Linear
Finite-Element Eddy-Current Problems via RL-Ladder Circuits 231**
Ruth V. Sabariego and Johan Gyselinck
21.1   Introduction 231
21.2   TWP28: Frequency-Domain FE Identification 233
21.3   Ladder-Circuit Approximation 236
References 238

Part VI   Industrial Applications

**22   A Lagrangian Approach to the Simulation of a Constricted
Vacuum Arc in a Magnetic Field**   243
Massimiliano Cremonesi, Attilio Frangi, Kai Hencken,
Marcelo Buffoni, Markus Abplanalp, and Jörg Ostrowski
22.1   Introduction 243
22.2   Minimal Arc Model 245
22.3   Numerical Method 247
    22.3.1   Boundary Conditions 248
22.4   Simplified Arc Test and Numerical Results 248
References 252

23   Virtual High Voltage Lab   255
Andreas Blaszczyk, Jonas Ekeberg, Sergey Pancheshnyi,
and Magne Saxegaard
23.1   Introduction 255
23.2   Basic Concept 256
23.3   VHVLab Components 257
    23.3.1   Gas Database 257
    23.3.2   Critical Spot Evaluation 258
    23.3.3   Evaluation of Discharge Path and Its Characteristics 258
    23.3.4   A 3-Stage Surface Charge Evaluation Procedure 259
23.4   Application Example: Ring Main Unit 259
23.5   Surface Charging 261
    23.5.1   Saturation Charge Formulation 261
    23.5.2   Case Study: Rod-Barrier Arrangement 261
23.6   Conclusion 263
References 263

Index 265

Contributors

Markus Abplanalp ABB Switzerland Ltd., Corporate Research, Baden-Dättwil, Switzerland

Sebastian Amsuess Otto Bock Healthcare Products GmbH, Research and Development, Vienna, Austria

Nicodemus Banagaaya Max Planck Institute for Dynamics of Complex Technical Systems, Magdeburg, Germany

Andreas Bartel Bergische Universität Wuppertal, Applied Mathematics and Numerical Analysis, Wuppertal, Germany

Werner Baumgartner Institute for Biomedical Mechatronics, Johannes Kepler University Linz, Linz, Austria

Theo G. J. Beelen Eindhoven University of Technology, Eindhoven, The Netherlands

Peter Benner Max Planck Institute for Dynamics of Complex Technical Systems, Magdeburg, Germany

Kai Bittner University of Applied Sciences of Upper Austria, Hagenberg, Austria

Andreas Blaszczyk ABB Corporate Research, Baden-Dättwil, Switzerland

Hans-Georg Brachtendorf University of Applied Sciences of Upper Austria, Hagenberg, Austria

Marcelo Buffoni ABB Switzerland Ltd., Corporate Research, Baden, Switzerland

Raffael Casagrande Seminar for Applied Mathematics, ETH Zürich, Zürich, Switzerland

Markus Clemens University of Wuppertal, Wuppertal, Germany

Massimiliano Cremonesi Politecnico di Milano, Milano, Italy

Herbert De Gersem TU Darmstadt, Darmstadt, Germany

Christian Diskus Institute for Microelectronics and Microsensors, Johannes Kepler University Linz, Linz, Austria

Jos J. Dohmen NXP Semiconductors, Eindhoven, The Netherlands

Andreas Dreizler Technische Universität Darmstadt, Fachgebiet für Reaktive Strömungen und Messtechnik, Center of Smart Interfaces, Darmstadt, Germany

Jennifer Dutiné University of Wuppertal, Wuppertal, Germany

Jonas Ekeberg ABB Corporate Research, Baden-Dättwil, Switzerland

Lihong Feng Max Planck Institute for Dynamics of Complex Technical Systems, Magdeburg, Germany

Attilio Frangi Politecnico di Milano, Milano, Italy

Peter Gangl Institut für Numerische Mathematik, TU Graz, Graz, Austria

Linz Center of Mechatronics GmbH (LCM), Linz, Austria

Kai Gausling Bergische Universität Wuppertal, Applied Mathematics and Numerical Analysis, Wuppertal, Germany

Renaud Gillon ON Semiconductor Belgium, Oudenaarde, Belgium

Erion Gjonaj Technische Universität Darmstadt, Institut für Theorie Elektromagnetischer Felder, Darmstadt, Germany

Michael Günther Bergische Universität Wuppertal, Wuppertal, Germany

Johan Gyselinck Université Libre de Bruxelles, BEAMS Dept., Brussels, Belgium

Christoph Hachtel Bergische Universität Wuppertal, Wuppertal, Germany

Md. Rokibul Hasan KU Leuven, Dept. Electrical Engineering (ESAT), Leuven, Belgium

EnergyVille, Genk, Belgium

Kai Hencken ABB Switzerland Ltd., Corporate Research, Baden, Switzerland

Thomas Henneron Laboratoire d'Electrotechnique et d'Electronique de Puissance, Arts et Metiers ParisTech, Lille, France

Ralf Hiptmair Seminar for Applied Mathematics, ETH Zürich, Zürich, Switzerland

Rick Janssen NXP Semiconductors, Eindhoven, The Netherlands

Peer W. Kämmerer Department of Oral, Maxillofacial and Plastic Surgery, University Medical Centre Rostock, Rostock, Germany

Eric R. Keiter Electrical Models and Simulation, Sandia National Laboratories, Albuquerque, NM, USA

Johanna Kerler-Back Universität Augsburg, Institut für Mathematik, Augsburg, Germany

Andreas Klaedtke Robert Bosch GmbH - Corporate Research, Robert-Bosch-Campus, Renningen, Germany

Grazia Lamanna Universität Stuttgart, Institut für Thermodynamik der Luft- und Raumfahrt, Stuttgart, Germany

Ulrich Langer Institute of Computational Mathematics (NuMa), JKU Linz, Linz, Austria

Qingzhe Liu Institut für Mathematik und Informatik, Ernst-Moritz-Arndt-Universität Greifswald, Greifswald, Germany

Sabrina Mairhofer Institute for Biomedical Mechatronics, Johannes Kepler University Linz, Linz, Austria

Peter Meuris Magwel NV, Leuven, Belgium

Laurent Montier Laboratoire d'Electrotechnique et d'Electronique de Puissance, Arts et Metiers ParisTech, Lille, France

Philipp-Thomas Müller RWTH Aachen, Aachen, Germany

Jan Niehof NXP Semiconductors, Eindhoven, The Netherlands

Jörg Ostrowski ABB Switzerland Ltd., Corporate Research, Baden, Switzerland

Yun Ouédraogo Technische Universität Darmstadt, Institut für Theorie Elektromagnetischer Felder, Darmstadt, Germany

Sergey Pancheshnyi ABB Zurich, Zurich, Switzerland

Carsten Potratz Robert Bosch GmbH - Corporate Research, Renningen, Germany

Andreas Preusche Technische Universität Darmstadt, Fachgebiet für Reaktive Strömungen und Messtechnik, Center of Smart Interfaces, Darmstadt, Germany

Roland Pulch Institute for Mathematics and Computer Science, Ernst-Moritz-Arndt-Universität Greifswald, Greifswald, Germany

Piotr Putek Bergische Universität Wuppertal, Wuppertal, Germany

Hendrikje Raben Institute of General Electrical Engineering, University of Rostock, Rostock, Germany

Theresa Roland Institute for Biomedical Mechatronics, Johannes Kepler University Linz, Linz, Austria

Wolfgang Roland Institute of Polymer Extrusion and Compounding, Johannes Kepler University Linz, Linz, Austria

Michael Friedrich Russold Otto Bock Healthcare Products GmbH, Research and Development, Vienna, Austria

Ruth V. Sabariego KU Leuven, Dept. Electrical Engineering (ESAT), Leuven, Belgium

EnergyVille, Genk, Belgium

Magne Saxegaard ABB Distribution, Skien, Norway

Thomas Werder Schläpfer ABB Switzerland Ltd., Corporate Research, Baden, Switzerland

Wim Schoenmaker Magwel NV, Leuven, Belgium

Sebastian Schöps Technische Universität Darmstadt, Graduate School CE, Darmstadt, Germany

Sebastian Schuhmacher Robert Bosch GmbH - Corporate Research, Renningen, Germany

Christoph Steinhausen Universität Stuttgart, Institut für Thermodynamik der Luft- und Raumfahrt, Stuttgart, Germany

Christian Strohm Dept. of Mathematics, Humboldt-University of Berlin, Berlin, Germany

Tatjana Stykel Universität Augsburg, Institut für Mathematik, Augsburg, Germany

Laura P. Swiler Optimization and Uncertainty Quantification, Sandia National Laboratories, Albuquerque, NM, USA

Bratislav Tasić NXP Semiconductors, Eindhoven, The Netherlands

E. Jan W. ter Maten Bergische Universität Wuppertal, Wuppertal, Germany

Caren Tischendorf Dept. of Mathematics, Humboldt-University of Berlin, Berlin, Germany

Ursula van Rienen Institute of General Electrical Engineering, University of Rostock, Rostock, Germany

Bernhard Weigand Universität Stuttgart, Institut für Thermodynamik der Luft- und Raumfahrt, Stuttgart, Germany

Thomas Weiland Technische Universität Darmstadt, Institut für Theorie Elektromagnetischer Felder, Darmstadt, Germany

Ian Z. Wilcox Component and Systems Analysis, Sandia National Laboratories, Albuquerque, NM, USA

Christoph Winkelmann ABB Switzerland Ltd., Corporate Research, Baden, Switzerland

Christoph Wolf Institute for Biomedical Mechatronics, Johannes Kepler University Linz, Linz, Austria

Ulf Zimmermann Institute of General Electrical Engineering, University of Rostock, Rostock, Germany

Acronyms

AC	Alternating Current
AD	Automatic Differentiation
AMF	Alternating Magnetic Field
BDF	Backward Difference Formulas
BDSM-ET	Block-diagonal structured model order reduction method for electro-thermal coupled problems with many inputs
BEM	Boundary Element Method
CAD	Computer-Aided Design
CB	Circuit Breakers
CFD	Computational Fluid Dynamics
CSPE	Cascaded Subspace Projection Extrapolation
CT	Computer Tomography
DAE	Differential-Algebraic Equation
DC	Direct Current
DG	Discontinuous Galerkin
DOF	Degrees Of Freedom
ECR	Electric Contact Resistances
EEC	Equivalent Electric Circuit
EMC	ElectroMagnetic Compatibility
EMG	ElectroMyoGraphy
EQS	Electro-QuasiStatic approximation
FD	Finite Difference
FDTD	Finite Difference Time Domain
FEM	Finite Element Method
GGD	Generalized Gaussian Density
GMRES	Generalized Minimal Residual Method
IBC	Impedance Boundary Conditions
IC	Integrated Circuit
KDE	Kernel Density Estimation
MC	Monte Carlo
MIMO	Multiple-Input-Multiple-Output

MNA	Modified Nodal Analysis
MOR	Model Order Reduction
MOSFET	Metal Oxide Semiconductor Field-Effect Transistor
MQS	Magneto-QuasiStatic approximation
MRI	Magnet Resonance Imaging
MSE	Mean Square Error
MVP	Magnetodynamic Vector Potential formulation
NOFE	Number Of used Function Evaluations
NURBS	Non-Uniform Rational B-Spline
NWIP	Non-Symmetric Weighted Interior Penalty
ODE	Ordinary Differential Equation
PC	Polynomial Chaos
PCE	Polynomial Chaos Expansion
PCG	Preconditioned Conjugate Gradient
PDE	Partial Differential Equations
PDF	Probability Density Function
PEEC	Partial Element Equivalent Circuit
PLL	Phase-Locked Loop
POD	Proper Orthogonal Decomposition
PSS	Periodic Steady State problems
QMC	Quasi Monte Carlo
RF	Radio Frequency
RFIC	Radio Frequency Integrated Circuit
RMU	Ring Main Unit
ROM	Reduced-Order Model
SCM	Stochastic Collocation Method
SDAE	Stochastic Differential Algebraic Equations
SISO	Single-Input-Single-Output
SPE	Subspace Projection Extrapolation
SVD	Singular Value Decomposition
TMF	Transverse Magnetic Field
UQ	Uncertainty Quantification
VCM	Volume Conductor Model
VCO	Voltage Controlled Oscillators
VHVLab	Virtual High Voltage Lab
XFEM	eXtended Finite Element Method

Part I
Computational Electromagnetics

Today Computational Electromagnetics (CEM) is one of the most established computational sciences, with applications in a broad range of areas. The development and design of electromagnetic devices is based on numerical simulation and optimization. The growing e-mobility has forced the industry to develop more efficient electrical motors. Moreover, CEM plays an important part in the life sciences. Here, we are often faced with highly complex mathematical models in which the electromagnetic fields interact with other fields. The simulation of the human heart is one such example. CEM is based on Maxwell's equations, which are composed of eight coupled Partial Differential Equations (PDEs) called Gauss' electric and magnetic laws, Faraday's law of induction, and Ampère's law with Maxwell's famous addition. Though these field equations were completed by James Clerk Maxwell (1831–1879) more than 150 years ago, they still provide challenging topics in the context of mathematical analysis, numerical analysis, scientific computing, and electrical engineering.

We assigned five papers to the CEM part of these proceedings, although most of the contributions of this book can be assigned to CEM:

U. van Rienen et al. conduct a first study on the application of electrical stimulation for intrinsic activation of bone healing processes in critical size defects of the facial skeleton in their paper *"Preliminary Numerical Study on Electrical Stimulation at Alloplastic Reconstruction Plates of the Mandible"*. The finite element simulation is based on the CT data and the potential equation for reconstructing the electric field strength. This is another example of how CEM is applied in the life sciences.

The paper *"Evaluation of Capacitive EMG Sensor Geometries by Simulation and Measurement"* by T. Roland et al. also makes a contribution to the application of CEM in medicine. The finite element COMSOL simulation, combined with a MATLAB nodal analysis, enables an efficient optimization of prostheses-sensor geometry.

In the paper *"Stability Analysis of Electromagnetic Transient Simulations"*, W. Schoenmaker et al. present an analysis of the stability characteristics of the discretized Maxwell-Ampère equations that result from a finite integration of the potential formulation of Maxwell's equations.

S. Schuhmacher et al. propose an efficient method for sensitivity analysis that relates changes in circuit parameters to changes in 3D model parameters, as detailed in their contribution *"Sensitivity of Lumped Parameters to Geometry Changes in Finite Element Models"*.

The paper *"Electro-Thermal Simulations with Skin-Layers and Contacts"* by C. Winkelmann et al. proposes an adaptive finite element method for computing ohmic losses in conductors. In a second step, the steady state temperature distribution is computed using the commercial CFD solver ANSYS Fluent. This approach is used to compute the temperature distribution in a circuit breaker, and the simulation results are in good agreement with the experimental data.

Chapter 1
Preliminary Numerical Study on Electrical Stimulation at Alloplastic Reconstruction Plates of the Mandible

Ursula van Rienen, Ulf Zimmermann, Hendrikje Raben, and Peer W. Kämmerer

Abstract It is well known that external biophysical stimulation via application of electric currents enhances bone healing and restores its structural strength. However, it has not yet been applied to treat large bone defects. We conducted a first study on application of electrical stimulation for intrinsic activation of bone healing processes in critical size defects of the facial skeleton. Basing on CT images of a patient with a critical size defect, a volume conductor model has been set up in which the stimulation electrodes are integrated. The problem can be modelled as a stationary current problem. It is solved by the finite element method. Different stimulation sites are studied with respect to the desired therapeutic range of the electric field strength. In future works, the model shall be further refined. The long-range aim is a patient-specific simulation within the therapy planning.

1.1 Medical Background and Previous Work

The integrity of the bony mandible plays a prominent role for function and aesthetics in the facial area. If there is disruption of the mandibular continuity, mainly due to tumour resection but also by the therapy of pronounced infections and after traumatic events, the affected patients suffer significant functional limitations and aesthetic shortcomings of the orofacial system. In selected cases alloplastic, usually made of titanium, bridging plates are used together with appropriate fixing screws on the remaining residual bone, which are introduced after resection of the tumour

U. van Rienen (✉) · U. Zimmermann · H. Raben
Institute of General Electrical Engineering, University of Rostock, Rostock, Germany
e-mail: ursula.van-rienen@uni-rostock.de; ulf.zimmermann@uni-rostock.de;
hendrikje.raben@uni-rostock.de

P. W. Kämmerer
Department of Oral, Maxillofacial and Plastic Surgery, University Medical Centre Rostock, Rostock, Germany
e-mail: peer.kaemmerer@med.uni-rostock.de

© Springer International Publishing AG, part of Springer Nature 2018
U. Langer et al. (eds.), *Scientific Computing in Electrical Engineering*,
Mathematics in Industry 28, https://doi.org/10.1007/978-3-319-75538-0_1

and the mandible for reconstruction of mandibular continuity defects. The goal is to maintain the mandibular guidance and prevention of mandible deflection by unilateral extension by muscles and scars. However, during the healing process often complications occur such as fractures of the titanium plates or loosening of the fixing screws, so that the survival of this system lasts often only for one year [1]. Such complications always require replacement of the osteosynthesis system and thus a re-invasive procedure with increased mortality and morbidity (and mortality rate impairment). It would therefore be of high priority, to realise a better growth of bone cells to the fixation system, thus ensuring an implant with improved fit and great longevity. Especially in the elderly further arises the problem of general decline of bone mass, which additionally complicates the mounting of osteosynthesis systems.

On the other hand, it is well known that the application of electromagnetic fields can enhance bone growth and bone remodelling. Assumedly, the natural piezoelectricity of bone and streaming potentials in the trabecular bone structure play an important role. Since the 1950s, the piezoelectricity of bone is common knowledge [2]. However, the evidence of enhanced bone growth due to artificial electromagnetic fields was provided by Bassett et al. only as late as in 1974 [3]. They also suspected that the piezoelectric behaviour is one major influencing factor in natural bone growth. Pollack et al. [4] published an early anatomical model for streaming potentials in osteons after first experimental observations. A more actual view on the role of streaming potentials may be found in [5]. Since the 1980s, magnetically induced electrical stimulation is commonly used as an adjunct in treating complications in bony healing [6]. This study considers macroscopic simulation models and thus does not deal with the microscopic processes such as piezoelectricity or streaming potentials but rather with electric field distributions aiming for field strengths that empirically lead to bone regeneration.

Recently, an electro-stimulative hip revision system applying such magnetically induced electrical stimulation has been designed [7, 8] and tested in vitro [9]: Based on tomographic data, a numerical model has been created employing a homogenisation approach for the soft tissue in the hip region and for the trabecular (inner) part of the hip bone, respectively. A number of stimulation electrodes have been placed on top of a commonly used acetabular cup. Their exact number and positions have been optimised for certain stages of bone damage exploiting the superposition property of the fields. To complement that electrically stimulating acetabular cup, flat electrodes have been designed for the hip stem. The stimulation principle has been validated in vitro and in vivo as well, the latter in a rabbit model.

To enhance the patient's comfort, now the aim is to develop medical devices that employ direct electrical stimulation, making an external primary induction coil redundant. Here, the mandible is studied as bone to be stimulated.

In the animal mandible, during the first 14–90 days of healing, accelerated bone growth was shown when using implants made of gold wires connected to a 3-V microlithium cell [10]. Another study stated a similar effect when applying intermittent $10\,V_{p-p}$, $60\,kHz$ sine-wave signals during osseointegration of dental implants [11]. The authors concluded that minimal direct electrical current resulting from an electric field around the implant could stimulate bone formation and decrease the time of osseointegration. Even so, further studies considering such treatment of bone defects are lacking.

All these findings encourage incorporating electrical stimulation into the therapy of critical size defects of facial skeleton. It seems to be a suitable way to promote the bone growth close to the osteosynthesis plates by means of electrical stimulation. For this, patient-individual numerical simulation models are highly desirable for therapy planning. Thus, in our preliminary study we consider a mandible reconstructed from fibula bone, for which we compare two different models. In future, we aim for using common alloplastic reconstruction plates [12] for electrical stimulation devices.

1.2   Set-Up and Solution of the Bio-Electric Model

In the sequel, we will describe the macroscopic bio-electric modelling and simulation step by step—each first in general, following the nomenclature of [13], and then for our specific problem under study:

Anatomic Modelling comprises determining the tissue distribution and, if applicable, e.g. fibre orientation and other physiological quantities, but also possible implants already in place, from tomographical data like e.g. Computer Tomography (CT) or Magnet Resonance Imaging (MRI). The resulting data sets are typically volume-oriented and consist of several hundreds of slices comprising a regular raster of so-called voxels, i.e. hexahedral volumes, which in general have nowadays a sub-millimetre resolution. Spatial resolution and total picture volume (image resolution), temporal resolution (acquisition speed) and tissue differentiability (tissue contrast, artefacts, and noise) comprise central parameters and limitations. In context of designing an electrically stimulating implant, the anatomical model will typically be supplemented with an implant and its stimulation electrodes and insulators.

Basing on sectional images from CT data of a patient with a critical size defect of the mandible caused by cancer, we registered and segmented the slices using Materialise Mimics® version 19 (http://www.materialise.com) and thus built a CAD model for the subsequent use in an electromagnetic field simulation. Figure 1.1 displays exemplary CT slices of the full data set consisting of 302 slices, each with 512×512 voxels of dimension $0.327 \times 0.327 \times 0.4$ mm.

We employed the individual gray values for a tissue or material. For segmentation, first the gray scale ranges were defined for the individual substances (air, soft tissue, bone, metal). This yields an intermediate, coarsely segmented model. This model was subsequently reduced by removing artefacts and parts that are outside of the region of interest. Thus nearly all bones, the ears and other small items were removed to minimise the model size. Of all bone tissue, only the jaw bone was kept. Afterwards the inner tissue boundaries were smoothed. Figure 1.2(left) shows the three-dimensional (3D) data after registration plus segmentation.

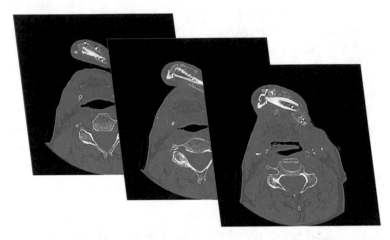

Fig. 1.1 Slices from computer tomography of a patient with a critical size defect of the mandible

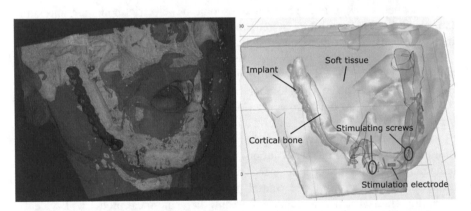

Fig. 1.2 (left) 3D data after registration and segmentation. This preliminary model differentiates between highly conductive titanium (red), conductive soft tissue (blue), and resistive bone (yellow). (right) CAD model. The metallic implant and the fixation screws are shown in blue. The two screws that are defined as field sources as well as the embedded stimulation electrode are highlighted. Note that only the relevant part of the bone (the mandible) is kept for the simulation model

As a further step, unwanted geometric features like spikes, self-intersections, and small components and holes where removed from the resulting STL object using the software GEOMAGIC Studio 12 (3D Systems, http://www.geomagic.com). After further smoothing, the polygon surface is converted into a NURBS surface which then can be easily imported into the simulation software.

Finally, Fig. 1.2(right) shows the anatomical 3D CAD model for which the so-called volume conductor model (VCM) [13] is set up. Note that ultimately the VCM, and thus the problem size, is further reduced such that only the mandible bone in a soft tissue background is modelled since this is sufficient to cover the region of interest, i.e. that volume where the field has not yet decayed to irrelevant size. That VCM serves as a representative model for our subsequent studies.

Physical Modelling constitutes determining tissue properties as well as the properties of the implant and assigning them to the anatomical model. The tissue properties depend on the specific tissue, fibre orientation, frequency, temperature, etc.. The tissue properties may be anisotropic or nonlinear, e.g. the electric conductivity could be nonlinear on the electric field. Also, they show large variations on individual and measurement, especially for low frequencies below 1 kHz [14]. These variations can be taken into account by Uncertainty Quantification, see e.g. [15]. Furthermore, the location of the field sources and their type of variation has to be specified.

We use the finite element (FEM) software COMSOL Multiphysics® version 5.2a (COMSOL, https://www.comsol.com) to set up the physical VCM from the anatomical CAD model. The electrical conductivity is taken from [14] and amounts to 7.4×10^5 S/m for titanium (Ti beta-21S), 0.7 S/m for soft tissue and 0.02 S/m for bone (cortical bone). We define sources using the implanted titanium prosthesis and screws as highlighted in Fig. 1.2b. According to the method of Kraus [6], we apply an alternating voltage of 20 Hz aiming at electric fields between 5 and 70 V/m. Below the threshold of 5 V/m, no beneficial effect to accelerated bone growth will be achieved, whereas electric fields above 70 V/m could damage adjacent soft tissue.

The next step involves the **set-up of the appropriate boundary value problem**. In bio-electric problems with low stimulation frequencies like the 20 Hz applied in this study, it is possible to simplify Maxwell's equations assuming that propagation and inductive effects are negligible [16–18]. This yields the so-called electro-quasistatic (EQS) approximation. Since the EQS field is free from eddy currents, it is uniquely defined by some scalar potential function. In time-harmonic case, the complex amplitude $\underline{\mathbf{E}}$ of the electric field strength can thus be expressed by a complex scalar potential $\underline{\varphi}$ via $\underline{\mathbf{E}} = -\nabla\underline{\varphi}$.

Plugging this into the EQS equations and assuming no impressed currents, implies the following Laplace equation

$$\nabla \cdot \Big([j\omega\varepsilon(\mathbf{r}) + \sigma(\mathbf{r})] \nabla\underline{\varphi}(\mathbf{r}) \Big) = 0 \qquad (1.1)$$

with the angular frequency ω, the electric permittivity $\varepsilon(\mathbf{r})$ and the electric conductivity $\sigma(\mathbf{r})$.

In case that further $\frac{\omega\varepsilon}{\sigma} \ll 1$, the problem can be thought of as a stationary current problem and

$$\nabla \cdot (\sigma(\mathbf{r})\nabla\varphi(\mathbf{r})) = 0 \qquad (1.2)$$

results, i.e. also the capacitive effects are neglected.

For the cortical bone involved in this preliminary study with $\varepsilon_r = 25,100$ and $\sigma = 0.02$ S/m the error in solving the stationary current problem (1.2) instead of the EQS problem (1.1) is estimated using the relation $\frac{\omega\varepsilon}{\sigma} = 0.001396$ to be below 0.14% at the given frequency of 20 Hz. Indeed, calculations comparing the electric field norm in the stationary and frequency domain studies showed the relative error to be below 0.14%. For this reason, we may well solve for the stationary current problem (1.2) instead of the EQS problem (1.1). The computation time reduces approximately by one third compared to an EQS approach.

Solution of the Boundary Value Problem We compute the electric potential $\varphi(\mathbf{r})$ in the bone tissue and at the implant surface solving the Laplace equation (1.2) in the computational domain. The boundary conditions are set to Dirichlet conditions with ±0.2 V at the field-inducing screws, to 0.4 V at the central electrode and to ground potential at the outer posterior boundaries of the bone where the metallic implant would be directly adjacent to the mandible, respectively. Neumann boundary condition has been applied for the outer shell of the soft tissue to represent the non-conductive air. We implied the Stationary Current Solver of COMSOL for the solution of this boundary value problem.

On basis of a mesh convergence study, the preliminary models used about 320,000 (screws) resp. 535,000 (electrode) tetrahedral mesh elements of an average quality 0.6 (a value of 1 would refer to equilateral mesh elements). This will be further improved in future works.

The resulting problem with 430,000 (screws) resp. 715,000 (electrode) degrees of freedom was solved with the conjugate gradient solver with a relative tolerance of $tol = 1 \times 10^{-3}$ and a factor $\rho = 400$ in COMSOL's error estimate to ensure the desired tolerance even for ill-conditioned problems. We used a Windows workstation with 24×3.00 GHz CPU and 256 GB RAM. The computation time was around 2 min per simulation run for both of the models described below allowing for extensive further parameter and optimisation studies in reasonable time, in future.

1.3 Comparison of Two Stimulation Sites

In this preliminary study, we compare two different stimulation sites. Firstly, in model 1, we allocate the stimulation site within the screws or nails that are anyhow involved in the implant. Details like cables are neglected in the simplified models described here. We use a stimulation voltage of 0.4 V.

Figure 1.3(left) displays a surface plot of the electric field norm on the mandible. It clearly elucidates that electric field strengths of 70 V/m and more are reached in the surrounding of the screws and therefore could lead to an undesired overstimulation of the adjacent soft tissue. Thus, a detailed parameter study will be done if smaller stimulation voltages would be applicable for this set-up to stay below the 70 V/m limit.

In a second case, in model 2, instead of using the screws as stimulation site we deliberately insert a stimulation electrode into the implanted bone. Specifically, we use a cylinder with diameter of 2 mm and height of 5 mm, see Fig. 1.2(right). Again we stimulate with 0.4 V. Figure 1.3(middle) similarly displays a surface plot of the

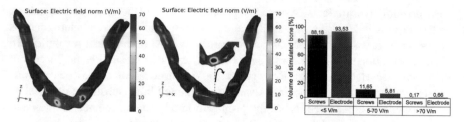

Fig. 1.3 (left) Model 1: surface plot of the electric field norm on the mandible in case of two screws as stimulation site. (middle) Model 2: surface plot of the electric field norm on the mandible for central stimulation site with implanted electrode. (right) Histogram showing the percentage of understimulated bone ($|\mathbf{E}| < 5$ V/m), optimally stimulated bone ($5 \leq |\mathbf{E}| \leq 70$ V/m), and overstimulated bone ($|\mathbf{E}| > 70$ V/m), respectively

electric field norm $|\mathbf{E}|$ on the mandible. Now, the maximum at the frontal surface stays below 35 V/m. Yet, Fig. 1.3(middle) reveals a broad region on the inner surface of the mandible where overstimulation with more than 70 V/m occurs.

To compare both stimulation sites, the percentage of beneficially stimulated bone is a good indicator. The histogram in Fig. 1.3(right) illustrates that the beneficially stimulated percentage of bone volume in the case with two stimulating screws (model 1) is more than twice the percentage of the electrode case (model 2). In addition, the (unwanted) overstimulated volume is slightly smaller in the screw configuration than in the electrode case. The non-stimulated volume is nearly the same in both cases. Thus, overall, the screw stimulation shows the better results. If imbedded stimulation electrodes would nevertheless be preferred, further studies should focus on the optimisation of the electrode position and stimulation voltage to ensure optimal stimulation in the defective region with electric field strengths between 5 and 70 V/m [15]. In any case, room for further improvement is expected in both models since this was only a preliminary study of first design ideas and a systematic parameter and optimisation study is pending. The optimisation will aim on the stimulation of the individual defective region. From that a region of interest will be defined. Analogous to [7], number and location of electrodes will be optimised for a maximal percentage of optimally stimulated bone regarding the region of interest. Next, we will also set up a numerical model that shall be compared with results from in vivo experiments, which are currently in preparation.

1.4   Conclusion and Outlook

Electromagnetic and electric stimulation is becoming a well-established method to enhance bone regeneration after surgeries. Our preliminary data shows that stimulation within facial skeleton defects can provide electric field strengths in the appropriate range to have a positive influence on the intrinsic activation of bone healing properties. These results are crucial for further investigations using electrostimulative implants. As next steps, the VCM will be enhanced by taking further technical details as well as a detailed differentiation between heterogeneous bone- and soft tissue into account and accordingly raising the mesh resolution. In close cooperation with the Department of Oral, Maxillofacial and Plastic Surgery at the University Medical Centre Rostock, various electrode geometries, localisations and other stimulation parameters as well as implanted bone substitute material as scaffold will be considered. Next, the propagation of parametric uncertainties will be accounted for using Uncertainty Quantification. In the long run, the simulation pipeline shall be carried over to an open source simulation tool implying higher-order FEM and being prepared for augmentation to a patient-individual therapy planning.

Acknowledgements This work was partly supported by the German Science Foundation (DFG) in the scope of the project DFG RI 814/24-1 and by a grant of the Federal State of Mecklenburg-Vorpommern.

References

1. Maurer, P., Eckert, A.W., Kriwalsky, M.S., Schubert, J.: Scope and limitations of methods of mandibular reconstruction: a long-term follow-up. Br. J. Oral Maxillofac. Surg. **48**, 100–104 (2010)
2. Fukada, E., Yasuda, I.: On the piezoelectric effect of bone. J. Phys. Soc. Jpn. **12**, 1158–1162 (1957)
3. Bassett, C.A., Pawluk, R.J., Pilla, A.A.: Acceleration of fracture repair by electromagnetic fields. a surgically noninvasive method. Ann. N. Y. Acad. Sci. **238**, 242–262 (1974)
4. Pollack, S.R., Petrov, N., Salzstein, R., Brankov, G., Blagoeva, R.: An anatomical model for streaming potentials in osteons. J. Biomech. **17**, 627–636 (1984)
5. Riddle, R.C., Donahue, H.J.: From streaming-potentials to shear stress: 25 years of bone cell mechanotransduction. J. Orthop. Res. **27**, 143–149 (2009)
6. Kraus, W.: Magnetic field therapy and magnetically induced electrostimulation in orthopedics. Orthopaede **13**, 78–92 (1984)
7. Potratz, C., Kluess, D., Ewald, H., van Rienen, U.: Multiobjective optimization of an electrostimulative acetabular revision system. IEEE Trans. Biomed. Eng. **57**, 460–468 (2010)
8. Klüß, D., Souffrant, R., Ewald, E., van Rienen, U., Bader, R., Mittelmeier, W.: Acetabuläre Hüftendoprothese mit einer Vorrichtung zur Elektrostimulation des Knochens. Patent (DE202008015661 U1) (2009)
9. Su, Y., Souffrant, R., Kluess, D., Ellenrieder, M., Mittelmeier, W., van Rienen, U., Bader, R.: Evaluation of electric field distribution in electromagnetic stimulation of human femoral head. Bioelectromagnetics **35**, 547–558 (2014)
10. Shayesteh, Y.S., Eslami, B., Dehghan, M.M., Vaziri, H., Alikhassi, M., Mangoli, A., Khojasteh, A.: The effect of a constant electrical field on osseointegration after immediate implantation in dog mandibles: a preliminary study. J. Prosthodont. **16**, 337–342 (2007)
11. Shigino, T., Ochi, M., Kagami, H., Sakaguchi, K., Nakade, O.: Application of capacitively coupled electric field enhances periimplant osteogenesis in the dog mandible. Int. J. Prosthodont. **13**, 365–372 (2000)
12. Kämmerer, P.W., Klein, M.O., Moergel, M., Gemmel, M., Draenert, G.F.: Local and systemic risk factors influencing the long-term success of angular stable alloplastic reconstruction plates of the mandible. J. Craniomaxillofac. Surg. **42**, e271–e276 (2014)

13. Malmivuo, J., Plonsey, R.: Bioelectromagnetism: Principles and Applications of Bioelectric and Biomagnetic Fields. Oxford University Press, New York (1995)
14. Gabriel, S., Lau, R.W., Gabriel, C.: The dielectric properties of biological tissues: II. Measurements in the frequency range 10 Hz to 20 GHz. Phys. Med. Biol. **41**, 2251–2269 (1996)
15. Schmidt, C., Zimmermann, U., van Rienen, U.: Modeling of an optimized electrostimulative hip revision system under consideration of uncertainty in the conductivity of bone tissue. IEEE J. Biomed. Health Inform. **19**, 1321–1330 (2015)
16. Plonsey, R., Heppner, D.B.: Considerations of quasi-stationarity in electrophysiological systems. Bull. Math. Biophys. **29**, 657–664 (1967)
17. van Rienen, U., Flehr, J., Schreiber, U., Motrescu, V.: Modeling and simulation of electro-quasistatic fields. In: Modeling, Simulation, and Optimization of Integrated Circuits. International Series of Numerical Mathematics, vol. 146, pp. 17–31. Birkhäuser, Basel (2003)
18. van Rienen, U., Flehr, J., Schreiber, U., Schulze, S., Gimsa, U., Baumann, W., Weiss, D., Gimsa, J., Benecke, R., Pau, H.W.: Electro-quasistatic simulations in bio-systems engineering and medical engineering. Adv. Radio Sci. **3**, 39–49 (2005)

Chapter 2
Evaluation of Capacitive EMG Sensor Geometries by Simulation and Measurement

Theresa Roland, Sabrina Mairhofer, Wolfgang Roland, Christian Diskus, Sebastian Amsuess, Michael Friedrich Russold, Christoph Wolf, and Werner Baumgartner

Abstract Myoelectric prostheses use electromyography (EMG) signals to control the movements of the prosthesis. EMG-signals are electric potentials on the skin which originate from voluntarily contracted muscles within a person's residual limb. Thus prostheses of this type utilize the residual neuro-muscular system of the human body to control the functions of an electrically powered prosthesis. Standard measurements are done using conductive electrodes on the skin surface. For technical reasons a capacitive coupling of the EMG to the prosthesis control would be preferable. To design optimal settings of the sensors, a detailed knowledge of the temporal electric potential distribution is vital. Here we show the simulation of the EMG using finite elements employing COMSOL based on MRI data. Then a node-based approach in MATLAB was derived and the comparison with the FE-results show that this approach yields excellent results and offers the advantage of high speed computation which allows for optimization of the sensor geometry. The simulation results were verified using measurements on volunteers showing that indeed our model assumptions and simplifications made are valid. The developed nodal analysis model enables fast and simple determination of the optimal prostheses-sensor geometry for the individual amputee.

T. Roland (✉) · S. Mairhofer · C. Wolf · W. Baumgartner
Institute for Biomedical Mechatronics, Johannes Kepler University Linz, Linz, Austria
e-mail: theresa.roland@jku.at; christoph.wolf@jku.at; werner.baumgartner@jku.at

W. Roland
Institute of Polymer Extrusion and Compounding, Johannes Kepler University Linz, Linz, Austria
e-mail: wolfgang.roland@jku.at

C. Diskus
Institute for Microelectronics and Microsensors, Johannes Kepler University Linz, Linz, Austria
e-mail: c.diskus@ieee.org

S. Amsuess · M. F. Russold
Otto Bock Healthcare Products GmbH, Research and Development, Vienna, Austria
e-mail: sebastian.amsuess@ottobock.com; michael.russold@ottobock.com

© Springer International Publishing AG, part of Springer Nature 2018
U. Langer et al. (eds.), *Scientific Computing in Electrical Engineering*,
Mathematics in Industry 28, https://doi.org/10.1007/978-3-319-75538-0_2

13

2.1   Introduction

Electromyography (EMG) measures the electrical signal generated by electrochemical effects in the muscles. It is used to control active, myoelectric prostheses. Amputees regain better quality of life and independence with these prostheses. In state-of-the-art EMG systems, conductive electrodes are used. This research deals with the approach to measure the EMG signal capacitively. In the capacitive measurement, no conductive connection between skin and sensor layer is necessary. This makes the system more independent of the skin condition. Pressure marks, which occur at conductive measurement, are avoided. This makes the system more applicable for people with circulation disorders. There are already some dealing with capacitive measurement of biosignals. Most of them deal with capacitive electrocardiogram, which has higher amplitudes compared to EMG (cf. [1] and [2]). Successful application of capacitive EMG signal to provide a new measurement system in prostheses is the aim of this research. This paper focuses on optimization of the sensor geometry by simulation and measurement.

2.2   Simulation of EMG

The simulation of the transmission of the action potential in the muscle fibre through the tissues is made by an FE model in COMSOL Multiphysics and with a nodal analysis in MATLAB. The tissue parameters are $\sigma_{Skin} = 1.14\,S/m$, $\varepsilon_{R\ Skin} = 2$, $\sigma_{Fat} = 457\,S/m$, $\varepsilon_{R\ Fat} = 20.81$, $\sigma_{Muscle} = 9329\,S/m$ and $\varepsilon_{R\ Muscle} = 266.71$ [3]. These parameters are used for both simulations.

2.2.1   Simulation in COMSOL Multiphysics

The COMSOL Multiphysics simulations are based on the results of Honeder's work [3]. An MRI image is segmented and used as a model for different tissues. The FE model consists of 86,585 elements. As a static signal source, the action potential described by Rosenfalck [4] is used. This potential, shown in Fig. 2.1, depends on the position z along the muscle fibre. The fibre is simulated at depth starting from 5 to 36 mm distance from the skin. The positions of the simulated muscle fibres are indicated in Fig. 2.2 by red dots. The resulting electric fields of those six muscle fibres are superpositioned in MATLAB, using the Helmholtz superposition principle. The muscle fibres at different depth are weighted with their circle segment area (the muscle is approximated to be circular). The circle used for the calculation is shown in Fig. 2.2 (red circle). The separation between two circle segments is at the midpoint between two neighboring muscle fibres.

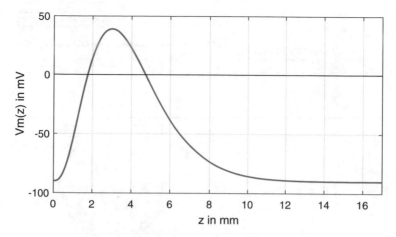

Fig. 2.1 Intracellular action potential V(z) along muscle fibre according to [4]

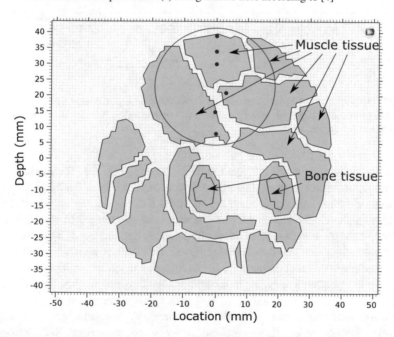

Fig. 2.2 MRI rendered human forearm [3]; simulated superposition muscle fibres (red dots); assumptive circular muscle for superposition (red circle); radius and ulna (bone tissue)

2.2.2   Simulation in MATLAB (Nodal Analysis)

The transmission of the action potential through the tissues in the human forearm is also simulated by nodal analysis. The basic principle of nodal analysis is explained in [5]. For this application, a three dimensional nodal analysis model was set up by

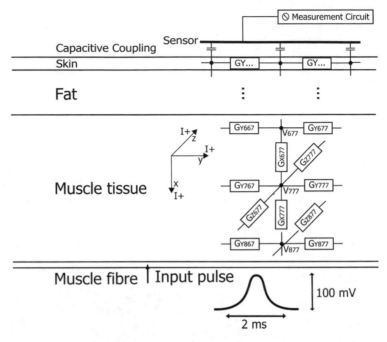

Fig. 2.3 Discreet 3D nodal analysis modelling forearm tissue layers; signal source, representing the action potential; snippet of voltage nodes and conductances; the signal is coupled to the measurement circuit via capacitances

the authors. A sketch of this discreet model is shown in Fig. 2.3. The model consists of homogenous tissue layers. Above the activated muscle fibre 10 mm muscle tissue, 4 mm fat and 2 mm skin are located. The nodal analysis model has eight discreet elements in the x-direction, which is the depth of the tissue. In the y-direction, longitudinal to the forearm, 80 discreet elements are applied. And in the z-direction, transverse to the forearm, 20 elements are used. This leads to a total of 12,800 nodes. The use of more discreet elements leads to a higher resolution of the electric field. The wave is smoother at a higher number of nodes, nevertheless, the shape basically remains the same. The choice of 12,800 nodes was determined empirically. The equidistant nodes are connected in all directions via these conductances. To achieve the voltages at the nodes, 2D conductance arrays were set up for each direction (Gx, Gy, Gz). The conductance arrays are automatically calculated, using the number of discreet elements and the dimensions in space. These 2D conductance arrays are transformed into a 3D matrix, which then is used for further calculation. In the transformation from 2D to 3D the model's edges are considered. This results in the 3D matrix G, containing the conductance values. With this matrix G a system of linear equations is set up (2.1). I_0 corresponds to the input current due to the action potential U_0 in the muscle fibre. G_0 is a 2D matrix containing the conductance values at the muscle tissue connecting the actuated muscle fibre. By transformation of the equation, the voltage values U can be calculated for each node. The voltages at the surface of the skin are of relevance for further evaluation. The action potential, represented by a 100 mV pulse, is applied at one discreet element. The action potential propagation speed is chosen to be 4 m/s in this model [6]. The action potentials are conducted with this propagation speed. The spatial distance between two action potentials is then defined by the action potential propagation speed and the action potential repetition rate. The typical EMG frequencies are in the range of 10–500 Hz. With a fatigued muscle lower frequencies (<100 Hz) dominate [7]. In this paper, an exemplary action potential repetition rate of 45 Hz is simulated in the nodal analysis. So after each 25 ms, another action potential with an amplitude of 100 mV is introduced. The effects of different action potential repetition rates and different tissue thicknesses are not focused on in this work.

$$\underline{\underline{G}} * \underline{U} + \underline{I_0} = \underline{0} \quad \longrightarrow \quad \underline{U} = -\underline{\underline{G}}^{-1} * \underline{I_0} \quad \underline{I_0} = \underline{U_0} * \underline{\underline{G}}_0^T \tag{2.1}$$

For the nodal analysis evaluation, the voltages at the nodes at the skin surface are used. The sensor is simulated with different areas, distances and with circular and rectangular geometry. The mean value of the skin surface voltage nodes located under the sensor is calculated as in the actual measurement. The measurement electronics with an amplifier gain of 1700 and the capacitive voltage divider at the input are considered in the simulation. The circuit input impedance is calculated with an input capacity of 16 pF. The nonlinear input resistance of the instrumentation amplifier is neglected. The coupling capacity is dependent on the sensor geometry. For the calculation of the coupling capacity, the ε_R of the insulation is 4, and a 0.8 mm sensor skin distance, due to the thickness of the insulation and a thin layer of dry skin, is used.

2.3 Experimental Methods

To compare the results of the simulations with real-world data, proband measurements were done and evaluated. Therefore the following measurement setup was used.

2.3.1 Measurement Setup

The sensors are made up of a flexible multi-layer construct of conducting and insulating material. Different layers are used for actual sensor and shielding. The sensor layers are separated from the skin by insulation as the EMG is measured capacitively. Two sensor layers are used to measure the differential signal. The measurement system developed by the research group is used for this work [2].

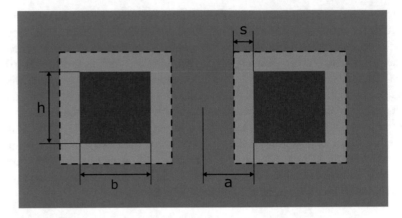

Fig. 2.4 Sensor geometry, (copper) sensor layer, (gray) active shield, (green) reference; *a* sensor layer distance to center, *b* sensor layer width/diameter, *h* sensor layer height, *s* active shield overhang

Table 2.1 Sensor geometries

Sensor	S1	S2	S3	S4	S5	S6	S7	S8
Sensor shape (rectangular (r), circular (c))	r	r	r	r	c	c	c	r
a (mm)	2.5	7.5	5	5	5	5	7.5	10
b (mm)	15	15	10	20	20	25	20	20
$A_{Sensorlayer}$ (mm^2)	300	300	200	400	314	490	314	400

Figure 2.4 shows a sketch of the sensor geometry. The parameters *a* and *b* are varied, *s* remains constant at 2 mm and *h* is 20 mm in the rectangular sensors. The EMG was measured with rectangular and circular sensor layers. The area of the reference should not influence the results, so the reference area is equal for every sensor. Table 2.1 lists the sensors used for the measurements. S5 equals the proposed geometry for conductive EMG according to SENIAM [8]. This range of sensor geometries is used due to practical reasons. The sensor size is limited by the physical size of the muscle group. It would not make sense to use a sensor larger than the muscle group. Same is valid for the sensor distance.

2.3.2 EMG Measurement

The measurements with eight different sensor geometries were done at seven probands. Each proband was measured at four contraction levels on the left and right forearm. The four contraction levels were relaxed muscle, forming a fist with maximum possible force, 10 and 15 kg at a 66fit spin grip hand training device. The sensor was positioned above the flexor digitorum superficialis muscle, at the center of the muscle belly. It was held in place with a stretchable stocking.

The measurement order was randomized, to prevent fatigue from influencing the evaluation results. At proband A, B and C order 1 (S9, S1, S2, S7, S8, S3, S4, S6) was chosen. For proband D, E, F and G order 2, the inverse of order 1, was chosen. The seven probands are in the range of 15–25 years old. The mean V_{RMS} value is used for the evaluation of the sensor geometries. For the calculation, a contraction sequence of 8 s at 10 kHz sampling frequency is used. A mean V_{RMS} value is calculated, comprising all measurements per sensor at the different contraction intensities at the left and right forearm of proband A-G. The V_{RMS} value of the measurements at relaxed muscle are subtracted to prevent noise from influencing the comparison of the sensors.

2.4   Results

The electric field at the skin surface, resulting from the superposition of the simulated muscle fibres in COMSOL Multiphysics, is shown in Fig. 2.5a. The results of the nodal analysis, simulating one action potential, are shown in Fig. 2.5b. The 3D representation of both simulations are cut along the length axis at forearm circumference 40 mm, see Fig. 2.5c. Differences in the amplitude and the broadness of the peaks are explained by different tissue models and superpositioning. In the FE simulation an MRI model is loaded, while in the nodal analysis a model of

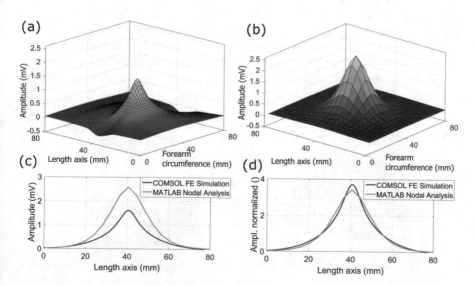

Fig. 2.5 (**a**) Electric field at skin surface determined by superposition of muscle fibres (COMSOL Multiphysics simulation); (**b**) electric field at skin surface resulting from one action potential determined by nodal analysis (MATLAB); (**c**) cut of 3D representation at forearm circumference 40 mm (FE simulation in COMSOL Multiphysics and nodal analysis in MATLAB); (**d**) normalized area of cut along length axis

homogenous tissue layers is applied. The quantitative difference in amplitude has no effect on the optimal sensor geometry, only the signal form is of importance (Fig. 2.5d). Figure 2.5d shows the cut along the length axis normalized to area. This normalization is done to compare the shape of the signals. For evaluation the absolute amplitude is of no importance for finding the optimum sensor geometry. The minor difference in shape are due to different tissue models applied. The relative difference between the two simulations results to 10%. The difference is calculated between the curves of the COMSOL and the MATLAB simulation. The area of the difference is calculated and referred to the area of the MATLAB Nodal Analysis curve. This difference is assessed by the authors as being acceptable for this application. In contrast to COMSOL, the nodal analysis model enables a simple implementation of the measurement with different sensor geometries. In COMSOL, the calculation of the resulting electric field at the skin surface from an action potential in one muscle fibre at one time step takes 22 s. In the nodal analysis in MATLAB, this calculation for one time step results to 0.23 s. Both computation times were determined on the same system (64 Bit Windows 7 Professional, Intel(R) Core(TM) i7-3770 CPU @ 3.40 GHz, 8 GB RAM). For this reason, the nodal analysis voltage values are used for further evaluation.

Figure 2.6 shows the results with an exemplary action potential repetition rate of 45 Hz. The measured amplitude at different sensor distances a is plotted over time in Fig. 2.6a. A sensor with an area of $400\,\mathrm{mm}^2$ is used in this plot. Due to the transmission through the tissue the action potential peak has an expansion at the skin surface. Therefore, at small sensor distances, the output voltage results to a smaller differential voltage, as it is measured by both sensor layers. If the sensor distance is too large and corresponding to the action potential repetition rate, the action potentials are cancelling each other out. The repetition rate and tissue thicknesses effect the damping and cancelling out of the action potentials and therefore determining the optimal distance. However, these two parameters

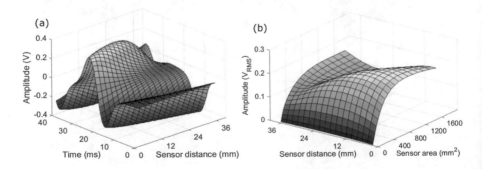

Fig. 2.6 Nodal analysis results; action potential repetition rate of 45 Hz. (**a**) Measured and amplified voltages at different sensor distances a over time at a rectangular sensor geometry with an area of $400\,\mathrm{mm}^2$; (**b**) V_{RMS} at different sensor areas and sensor distances for rectangular sensor geometry

remain constant in this evaluation, their impact will be studied in future work. Figure 2.6b plots the measured V_{RMS} over sensor area A and sensor distance a. The V_{RMS} is calculated using the measurement values over time. At small sensor areas, the amplitude increases with sensor area. This results from the capacitive voltage divider at the input. For this reason, the circuit impedance, respectively the input capacity, has to be minimized and the coupling capacity shall be increased e.g. by thinner insulation. At smaller circuit input capacities the V_{RMS} maximum would be shifted to smaller sensor areas. Due to a limited width of the action potential peak, the voltage is decreasing at sensor areas larger than the optimum. For this reason, it is essential to know the circuit input characteristics and to determine the optimal sensor area for the used measurement system. The optimal sensor geometry for measuring the action potentials with a 45 Hz repetition rate is calculated in MATLAB. It is resulting to a sensor distance a of 10 mm and a sensor area A of 676 mm^2. This maximum is calculated using the MATLAB function 'fmincon'. The costs which are minimized are the inverse of V_{RMS}. The inverse of the V_{RMS} is used for the minimization procedure, as we want to find the V_{RMS}'s maximum. There are constraints in sensor area and sensor distance, due to the limited size of the nodal analysis model. The sensor distance has a lower bound of 2 mm and an upper bound of 36 mm. The sensor area's lower bound is 4 mm^2 and the upper bound is 1600 mm^2. The starting point used for this optimization is equivalent to the lower bounds. The 'Global Search' method in MATLAB is used to prevent stopping at local extrema. This maximum found can be seen in Fig. 2.6b.

In Fig. 2.7 the V_{RMS} values of the proband measurements are plotted (S1–S8). The shape of the marker indicates whether the sensor geometry is circular or rectangular (cf. Table 2.1). The results of the nodal analysis simulation are plotted for different sensor distances and for circular and rectangular sensors over the sensor area. The signals are plotted for sensor distances in the range 2–10 mm according to real sensors used for the measurements. The unevenness in the curve of the circular geometry is explained by discretization of the circular measurement sensor element. The circular sensor shows slightly higher amplitudes at higher sensor areas, because its shape is similar to the action potential at the surface of the skin. At smaller areas, the difference between circular and rectangular geometry is dominated by the sensor area.

Exemplifying sensor 8 is plotted for the nodal analysis (S8$_{NA}$) and for the finite element model (S8$_{FE}$). A sensor element with the size of sensor 8 is placed above the action potential peak calculated by the nodal analysis and the finite element simulation. The mean value of the electric field at the skin surface beneath the sensor is calculated for both simulations. The calculation of the mean value represents the physical measurement, as it is explained in Sect. 2.2.2. As can be seen in Fig. 2.7, the nodal analysis (S8$_{NA}$) overestimated the signal amplitude by about 0.02 V while the FE-analysis (S8$_{FE}$) underestimated the amplitude of the physiological signal (S8) by about 0.01 V. The nodal analysis deviates from the FE-analysis by about 12%. Given the measured amplitude of 0.2 V and being aware, that a lot of physiological parameters are unknown or had to be estimated, the deviations of 10% of the areas of the 2D slice are tolerable (see Fig. 2.5d). This is true especially

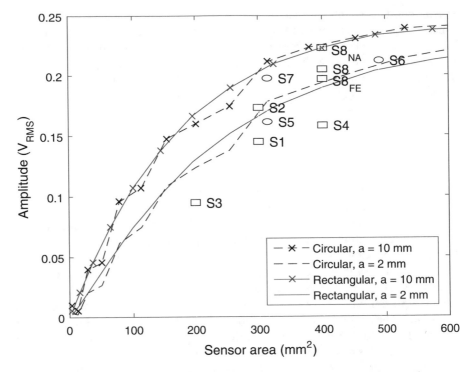

Fig. 2.7 Nodal analysis simulated V_{RMS} for circular and rectangular sensor geometry by sensor area A at different distances a; real-world measurement values for different sensor geometries (Table 2.1)

when considering that the absolute amplitude does not influence the optimal sensor geometry, only the shape of the electric field at the skin surface is doing so. Given the fact that the FE-analysis relies on NMR-data and a much more sophisticated and more time consuming mathematical approach, we are convinced that the simple and fast nodal analysis is well suited for the practical application of the design and personally individualized optimization of sensor geometries for different patients.

2.5 Conclusion

With this work, the geometry of capacitive sensors to achieve maximal signal coupling is determined. The distance between the sensor layers should be large enough to measure the differential voltage due to the action potential. This sensor layer distance should not correspond to the distance between the action potentials, as two action potentials would cancel each other out. Different EMG frequencies and different tissue thicknesses influence the damping and cancellation within the tissue, this will be investigated in future work. The sensor area should be chosen large enough to have sufficient signal coupling, but not too large to prevent capacities towards low voltage. The circular geometry should be preferred over the rectangular one, although this difference is minute. A small circuit input impedance leads to better signal coupling and the optimal sensor area shifts to smaller dimensions, which leads to better applicability in myoelectric prostheses. In future work, input capacity neutralization circuits will be investigated. The input circuit has to be designed in a way to reduce parasitic capacities. Active guarding plays an important role to do so. The nodal analysis is a very fast and simple approximation and its validity is verified by FE-simulation and measurements. The optimal sensor geometries can now be determined easily for the individual patient using this method.

Acknowledgements This work was supported by Otto Bock Healthcare GmbH and the Linz Center of Mechatronics.

References

1. Heuer, S.: Ambient capacitive ECG measurement-electrodes, systems and concepts ("Ambiente kapazitive EKG-Messung - Elektroden, Systeme und Konzepte"). Dissertation, Karlsruher Institute for Technology, Rottweil (2011)
2. Roland, T., Amsuess, S., Russold, M.F., Wolf, C., Baumgartner, W.: Capacitive sensing of surface EMG for upper limb prostheses control. Procedia Eng. **168**, 155–158 (2016). https://www.sciencedirect.com/science/article/pii/S1877705816335007
3. Honeder, J.L.: A quasi-stationary approach to the approximate solution of finite element analysis applied to EMG modeling. Master's thesis, Technikum Wien, Vienna (2013)
4. Rosenfalck, P.: Intra- and extracellular potential fields of active nerve and muscle fibres. Acta Phys. Scand. Suppl. **321**, 1–168 (1969)
5. Bakshi, U.A., Bakshi, A.V.: Circuit Analysis. Technical Publications, Pune (2007)
6. Andreassen, S., Arendt-Nielsen, L.: Muscle fibre conduction velocity in motor units of the human anterior tibial muscle: a new size principle parameter. J. Physiol. **391**, 561–571 (1987)
7. Bruns, T., Praun, N.: Biofeedback - Ein Handbuch für die therapeutische Praxis. Vadenhoeck & Ruprecht, Göttingen (2002)
8. Hermens, H.J., Freriks, B., Disselhorst-Klug, C., Rau, G.: Development of recommendations for SEMG sensors and sensor placement procedures. J. Electromyogr. Kinesiol. **10**, 361–374 (2000)

Chapter 3
Stability Analysis of Electromagnetic Transient Simulations

**Wim Schoenmaker, Christian Strohm, Kai Bittner,
Hans Georg Brachtendorf, and Caren Tischendorf**

Abstract We present an analysis of the stability characteristics of the discretized Maxwell-Ampere equations that result from a finite integration of the potential formulation. We demonstrate that the derivation of the discrete versions of these equations will result into unstable formulations unless, in the conversion from a continuous expression to a discrete expression, one accounts for the original motivation of the presence of the prior form.

3.1 Introduction

It is a well-known fact that electrical systems containing resistors will respond to transient signals in such a way that when the stimulus stops at some time instant the electromagnetic fields will gradually decay due to two physical mechanisms. First of all the resistances convert electrical energy into heat such that the electric energy decreases. Secondly for open systems there is radiation loss which also results into the situation that the electrical energy decays when time proceeds.

When constructing a transient simulation tool of electrical systems it is required that the basic fact of above energy decay mechanism is mimicked by the simulator. For circuit simulation tools this fact is easily reproduced because the circuit equations that contain resistors are in general stable. We can identify stability as a property of the circuit equations in the following way.

W. Schoenmaker (✉)
MAGWEL NV, Leuven, Belgium
e-mail: wim.schoenmaker@magwel.com

C. Strohm · C. Tischendorf
Dept. of Mathematics, Humboldt-University of Berlin, Berlin, Germany
e-mail: strohmch@math.hu-berlin.de; tischendorf@math.hu-berlin.de

K. Bittner · H. G. Brachtendorf
University of Applied Sciences of Upper Austria, Hagenberg, Austria
e-mail: Kai.Bittner@fh-hagenberg.at; Hans-Georg.Brachtendorf@fh-hagenberg.at

© Springer International Publishing AG, part of Springer Nature 2018
U. Langer et al. (eds.), *Scientific Computing in Electrical Engineering*,
Mathematics in Industry 28, https://doi.org/10.1007/978-3-319-75538-0_3

Let $\mathbf{X}(t)$ be the collection of all system or circuit variables. The complete system of circuit equations is given by the state-space equations

$$E \frac{d}{dt}\mathbf{X} + A\mathbf{X} = 0 \ . \tag{3.1}$$

If E is a non-singular matrix we may rewrite (3.1) as

$$\frac{d}{dt}\mathbf{X} + J\mathbf{X} = 0, \quad J = E^{-1}A \ . \tag{3.2}$$

Stability corresponds to the property of J that all its eigenvalues have real parts larger than or equal zero. If some eigenvalues have real part less than zero, the system has modes that explode when time proceeds which conflicts the energy conservation law and the system is therefore unphysical. Of course if it is impossible to create initial conditions such that when decomposed into the eigenvector base there are no components corresponding to negative real-part eigenvalues one may conclude that these modes will neither develop in the future and therefore the formulation is physical acceptable. Unfortunately this does not mean that if such a formulation of the system equations exists, e.g. J has negative real-part eigenvalues but the initial condition projected onto the negative real-part eigenvectors is empty, the simulation set up is physically save. While the transient time steps accumulate, numerical noise can mix into the transient solution and after some time leap the solution can still explode and yet becomes physically unacceptable. This was nicely demonstrated in [1].

As is seen from (3.2), the stability criterion could be straightforwardly connected to the formal solution

$$\mathbf{X}(t) = \mathbf{X}(0)e^{-Jt} \ . \tag{3.3}$$

This solution is easily obtained because the system equations are *first* order in time. When the Maxwell-Ampere equations are considered we must account for the wave-like solutions and these equations are *second* order in time. The stability analysis must be revised. The Maxwell-Ampere equations in the potential formulation are given in [2]. By introducing the variable $\boldsymbol{\Pi} = \partial_t\mathbf{A}$ the second-order system of equations is converted to first-order. Of course this step does not change the characteristic features of the solution, but it makes the system accessible to regular stability analysis.

3.2 Discretization Procedure for the Maxwell-Ampere Equation

The four Maxwell's equations read

$$\nabla \times \mathbf{E} = -\partial_t\mathbf{B}, \quad \nabla \times \mathbf{H} = \mathbf{J} + \partial_t\mathbf{D}, \quad \nabla \cdot \mathbf{D} = \rho, \quad \nabla \cdot \mathbf{B} = 0 \ . \tag{3.4}$$

where $\mathbf{E}$, $\mathbf{D}$ are the electric field strength and the displacement, and $\mathbf{H}$, $\mathbf{B}$ the magnetic field strength and induction, respectively. Moreover ρ and $\mathbf{J}$ are the electric charge density and current density, respectively. In what follows we assume linear isotropic materials, i.e.

$$\mathbf{D} = \epsilon\,\mathbf{E}, \quad \mathbf{B} = \mu\,\mathbf{H}, \quad \mathbf{E} = -(\nabla V + \partial_t \mathbf{A}), \quad \mathbf{B} = \nabla \times \mathbf{A} \ . \tag{3.5}$$

where ϵ is the dielectric constant and μ the permeability. It should be noted that materials of different types can be stacked or blocks of different materials can be placed next to each other. This results into abrupt jumps in the overall permittivity ϵ and permeability μ. However, generally we assume that the parameters depend on the space coordinate. Furthermore, we rewrite the Maxwell equations using the scalar potential V and vector potential $\mathbf{A}$.

3.2.1 Spatial Discretization

In what follows we introduce the following notation. For numbering the space grid points of adjacent grid nodes we use i and j. Moreover, let Δw_i be a finite volume element, associated with a grid node i, and $\sigma_{ij} = \pm 1$ the orientation of a link connecting node i and node j. It is set positive when oriented from inside to outside of the volume. The links between nodes i and j are denoted with $\langle ij \rangle$ and have an associated length h_{ij} and an area ΔS_{ij} for the dual surface. Every link has an *intrinsic* orientation vector of length 1 and is denoted by $\mathbf{e}_{ij}$. The projection of the vector potential onto a link $\langle ij \rangle$ is marked with index i and j, e.g. $\mathbf{e}_{ij} \cdot \mathbf{A} = A_{ij}$. When applying the finite-volume method, each node generates a balance equation corresponding to elaborating the divergence of a flux over the surface of the dual volume element of each node. Each surface element naturally gets a normal vector pointing away from the node under consideration. This vector $\mathbf{n}$, that is also found on each link, can be parallel or anti-parallel to $\mathbf{e}$. The resulting sign is denoted as s, e.g. $s_{ij} = \mathbf{n} \cdot \mathbf{e} = \pm 1$. The discretization of Gauss law is done using the usual finite-volume method or finite-integration technique.

The Maxwell-Ampère law is addressed in a slightly different way. Let $\mathbf{J}_c$ be the conduction current. It reads

$$\frac{1}{\mu} \nabla \times \nabla \times \mathbf{A} = \mathbf{J}_c - \epsilon \frac{\partial}{\partial t} (\nabla V + \mathbf{\Pi}) \ . \tag{3.6}$$

In order to obtain a unique solution we must impose a gauge condition. Here we use

$$\frac{1}{\mu} \nabla (\nabla \cdot \mathbf{A}) + \xi \varepsilon \nabla \left(\frac{\partial}{\partial t} V \right) = 0 \ , \tag{3.7}$$

where $0 \leq \xi \leq 1$ is a free parameter. The Coulomb gauge is obtained with $\xi = 0$ and the Lorenz gauge with $\xi = 1$ as special cases. Adding the gauge condition (i.e. zero to the right-hand side of (3.6), performing an integration over a dual-surface area ΔS of a link $\langle ij \rangle$ and multiplying the result with the length $L = h_{ij}$ of the link under consideration gives with $\mathbf{J}_c = \sigma \mathbf{E}$

$$
\begin{aligned}
\epsilon L \frac{\partial}{\partial t} \int_{\Delta S} d\mathbf{S} \cdot \mathbf{\Pi} = {} & -L \int_{\Delta S} d\mathbf{S} \cdot \nabla \times \left(\frac{1}{\mu} \nabla \times \mathbf{A} \right) + L \int_{\Delta S} d\mathbf{S} \cdot \frac{1}{\mu} \nabla (\nabla \cdot \mathbf{A}) \\
& - L \int_{\Delta S} d\mathbf{S} \cdot \sigma \nabla V - L \int_{\Delta S} d\mathbf{S} \cdot \sigma \mathbf{\Pi} \\
& - \epsilon L \int_{\Delta S} d\mathbf{S} \cdot \frac{\partial}{\partial t} (\nabla V) + \xi \epsilon L \int_{\Delta S} d\mathbf{S} \cdot \nabla \left(\frac{\partial}{\partial t} V \right) .
\end{aligned}
\tag{3.8}
$$

The discretization of each term will now be discussed. Starting at the left-hand side, we define a link variable Π_{ij} for the link going from node i to node j. The surface integral is approximated by taking $\mathbf{\Pi}$ constant over the dual area. Thus

$$
\epsilon L \frac{\partial}{\partial t} \int_{\Delta S} d\mathbf{S} \cdot \mathbf{\Pi} \simeq \epsilon L \, \Delta S_{ij} \frac{d\Pi_{ij}}{dt} .
\tag{3.9}
$$

We can assign to each link a volume being $\Delta v_{ij} = L \, \Delta S_{ij}$.

Remark 3.1 Note that $\Delta v_{ij} \neq \Delta w_{ij}$, since Δv_{ij} is the volume corresponding to the area of a dual surface multiplied with the length of a primary-mesh link whereas Δw_{ij} is a dual volume of a primary-mesh node.

The first term on the right-hand side is dealt with using Stokes theorem twice in order to evaluate the circulations

$$
-L \int_{\Delta S} d\mathbf{S} \cdot \nabla \times \left(\frac{1}{\mu} \nabla \times \mathbf{A} \right) = -L \oint_{\partial(\Delta S)} d\mathbf{l} \cdot \left(\frac{1}{\mu} \nabla \times \mathbf{A} \right) .
\tag{3.10}
$$

The circumference $\partial(\Delta S)$ consists of N segments. Each segment corresponds to a dual link that pierces through a *primary*-mesh surface. Therefore, we may approximate the right-hand side of (3.10) as

$$
-L \oint_{\partial(\Delta S)} d\mathbf{l} \cdot \left(\frac{1}{\mu} \nabla \times \mathbf{A} \right) = -L \sum_{k=1}^{N} \Delta l_k \frac{1}{\mu_k} (\nabla \times \mathbf{A})_k ,
\tag{3.11}
$$

where the sum goes over all primary-mesh surfaces that were identified above as belonging to the circulation around the starting link. Note that we also attached an index on μ. This will guarantee that the correct value is taken depending in which material the segment Δl_k is located. Next we must obtain an appropriate expression for $(\nabla \times \mathbf{A})_k$. For that purpose, we consider the primary-mesh surfaces. In particular, an approximation for this expression is found by using

$$(\nabla \times \mathbf{A})_k \simeq \frac{1}{\Delta S_k} \int_{\Delta S_k} d\mathbf{S} \cdot \nabla \times \mathbf{A} = \frac{1}{\Delta S_k} \oint_{\partial(\Delta S_k)} d\mathbf{l} \cdot \mathbf{A} \ . \tag{3.12}$$

The last contour integral is evidently replaced by the collection of primary-mesh links variables around the primary-mesh surface. As a consequence, the first term at the right-hand side of (3.8) becomes

$$- L \sum_{k=1}^{N} \Delta l_k \frac{1}{\mu_k} \frac{1}{\Delta S_k} \left(\sum_{l=1}^{N'} \Delta l_{\langle kl \rangle} A_{\langle kl \rangle} \right) \ , \tag{3.13}$$

where we distinguished the link labeling from node labeling (ij) to surface labeling $\langle kl \rangle$.

Next we consider the second term of (3.8). Now we use the fact that each link has a specific *intrinsic* orientation from 'front' to 'back' that was earlier set equal to $\mathbf{e}$,

$$L \int_{\Delta S} d\mathbf{S} \cdot \frac{1}{\mu} \nabla (\nabla \cdot \mathbf{A}) \simeq \int_{\Delta S} d\mathbf{S} \cdot \frac{1}{\mu} (\nabla \cdot \mathbf{A})_{back} - \int_{\Delta S} d\mathbf{S} \cdot \frac{1}{\mu} (\nabla \cdot \mathbf{A})_{front} \ . \tag{3.14}$$

The two terms in (3.14) are now discretized as

$$\int_{\Delta S} d\mathbf{S} \cdot \frac{1}{\mu} (\nabla \cdot \mathbf{A}) = \frac{\Delta S}{\mu \Delta v} \int_{\Delta v} dv \nabla \cdot \mathbf{A} = \frac{\Delta S}{\mu \Delta v} \oint_{\partial(\Delta v)} d\mathbf{S} \cdot \mathbf{A} = \frac{\Delta S}{\mu \Delta v} \sum_{j}^{n} \Delta S_{ij} A_{ij} \ ,$$

where the sum is now from the front or back node to their corresponding neighbor nodes. The boundary conditions enter this analysis in a specific way. Suppose the front or back node is on the surface of the simulation domain. Then the closed surface integral around such a node will require a dual area contribution from a dual area outside the simulation domain. These surfaces are by definition not considered. However, we can return to the gauge condition and use

$$\int_{\Delta S} d\mathbf{S} \cdot \frac{1}{\mu} (\nabla \cdot \mathbf{A}) = -\xi \Delta S \, \epsilon \frac{\partial V}{\partial t}. \tag{3.15}$$

At first sight this looks weird: First we insert the gauge condition to get rid of the singular character of the curl-curl operation and now we 'undo' this for nodes at the surface. This is however fine because for the Dirichlet boundary conditions for $\mathbf{A}$ there are no closed circulations around primary surfaces and there is no uniqueness problem and therefore the double circulation operator is well defined.

The next two terms are rather straightforward: For the third term we consider ∇V constant over the dual surface. Thus we obtain

$$- L \int_{\Delta S} d\mathbf{S} \cdot \sigma \nabla V = (V_{front} - V_{back}) \left(\sum \Delta S_i \sigma_i \right) . \tag{3.16}$$

The variation of σ is taken into account by looking at each volume contribution separately. The fourth term can be dealt with in a similar manner

$$- L \int_{\Delta S} d\mathbf{S} \cdot \sigma \mathbf{\Pi} = L \, \Pi_{ij} \left(\sum \Delta S_i \sigma_i \right) . \tag{3.17}$$

Of critical importance are the details of the implementation of the gauge condition. In order to make the double-curl operator more Laplacian-like we added the gauge condition to this equation. There are two terms that contain a mixture of a spatial and a time derivative, i.e.

$$- \frac{\partial}{\partial t} (\epsilon \nabla V) \quad \text{and} \quad \xi \nabla \left(\epsilon \frac{\partial V}{\partial t} \right) . \tag{3.18}$$

It turns out that these terms need a different discretization based on the origin of appearance in the Maxwell-Ampere equation. The first term in (3.18) needs to be discretized as is done for the term $\frac{\partial}{\partial t} (-\mu_0 \epsilon \mathbf{\Pi})$. The discretization is based on the finite-*surface* integration, whereas the second term in (3.18) needs to be discretized as is done for the term $\nabla(\nabla \cdot \mathbf{A})$. The latter is discretized using the finite-*volume* discretization. We observed that dealing with both terms using the finite-volume discretization leads to an unstable discretized formulation of the Maxwell-Ampere system. This is demonstrated in the following numerical example. In Fig. 3.1, a

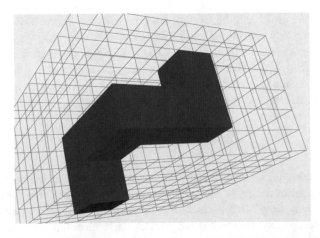

Fig. 3.1 Twisted bar used for computing the spectrum of the matrix that determines the system stability

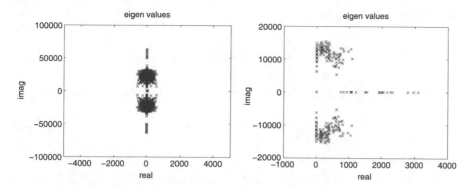

Fig. 3.2 Zoom-in to the eigenvalue spectrum around the real axis using exclusively finite-volume discretization for terms containing a mixed temporal and spatial differentiation (left). Zoom-in to the eigenvalue spectrum around the real axis using finite-volume discretization and finite-surface integration for terms containing a mixed temporal and spatial differentiation (right)

twisted bar is shown and a coarse mesh is used. This allows us to do a detailed eigenvalue analysis of the discretized system. In Fig. 3.2 left panel, the spectrum of the matrix J is shown based on a finite-volume implementation of both terms in (3.8). In Fig. 3.2 right panel, the spectrum of J is shown where J results from a discretization of (3.18) using the finite-*surface* integration method for the left term and keeping the discretization of the right term unaltered. We demonstrated that the conversion of continuous terms to discrete representative terms must account for the original motivation behind their presence.

3.3 Some Theoretical Considerations

Let us return to the Maxwell-Ampere equation (3.6). Since $\mathbf{J}_c = \sigma \left(-\nabla V - \mathbf{\Pi} \right)$ and $\mathbf{\Pi} = \partial_t \mathbf{A}$ we obtain

$$\epsilon \frac{\partial^2}{\partial t^2} \mathbf{A} + \sigma \frac{\partial}{\partial t} \mathbf{A} = M_{op} \mathbf{A} + \mathbf{J}_s \, , \tag{3.19}$$

$$M_{op} = \nabla \left(\nabla \cdot \left(\frac{1}{\mu} \right) \right) - \nabla \times \left(\frac{1}{\mu} \nabla \times \right) \simeq \left(\frac{1}{\mu} \right) \nabla^2 \, . \tag{3.20}$$

In here M_{op} is a spatial differential operator and $\mathbf{J}_s$ is a source term. For a planar structure, the component A_z decouples from the equations system. Moreover the source term for this component is zero. The second order spatial derivative will lead to wave-like solutions. Consider the very simple one-DOF equation and k^2 is the result of the Laplace operator:

$$\epsilon \frac{\partial^2 x}{\partial t^2} + \sigma \frac{\partial x}{\partial t} + k^2 x = 0 . \tag{3.21}$$

There are solutions of the type $x(t) = x_0 \exp(\lambda t)$. Inserting this solution gives:

$$\epsilon \lambda^2 + \sigma \lambda + k^2 = 0 . \tag{3.22}$$

The solutions of this equation for $\sigma > 0$ are:

$$\lambda_{1,2} = -\frac{\sigma}{2\epsilon} \left(1 \pm \sqrt{1 - \frac{4\epsilon k^2}{\sigma^2}} \right) . \tag{3.23}$$

This cannot lead to an unstable eigenvalue since the argument of the square root is a number less than one. If the argument is less than zero we get wave-like solutions. Our observation critically depends on the assumption that the 'Laplace operator' M_{op} gives rise to $k^2 \geq 0$. Unstable eigenvalues can arise if M_{op} gives rise to negative eigenvalues for the imposed boundary conditions. It should also be noted that if $\sigma = 0$ then the eigenvalues become pure imaginary.

3.4 The Impact of Meshing

In this section we consider a structure with contacts at the edge of the simulation domain. The structure and its mesh are shown in Fig. 3.3. The Manhattan meshing gives rise to an eigenvalue spectrum which has no negative real-parts. However,

Fig. 3.3 Test structure: 2D view (left) and 3D view (right)

when using $2D$—Delaunay meshing, we find that the spectrum has severe negative real-part eigenvalues: $(-1.92\ 10^{10}+0.0i), (-4.92+4.26\ 10^3i), (-4.92-4.26\ 10^3i)$. As is seen in Fig. 3.3 left panel, some cells have obtuse angles. This will lead to negative dual areas. One may modify the meshing algorithm by assigning a dual volume to each node in each cell by starting from the center of gravity for the surfaces of the cell and the cell volume. Using this modified method of obtaining dual volumes and dual areas, the negative real-part eigenvalues are removed again.

3.5 Conclusion

Converting the Maxwell-Ampere equations that are second-order in time differentiation into equations that are first-order in time differentiation, the standard techniques for stability consideration become applicable. We found that the discretization of each term must be done in accordance with the original motivation of the appearance of the term. Carelessly swapping temporal and spatial differential operators may quickly lead to erroneous discretization set up. We also noted that a stable implementation requires that the discrete Laplace operator must be implemented such that its continuous spectrum property, i.e. semi-definiteness must be preserved.

Acknowledgements This work is funded by the European FP7 project nanoCOPS under grant 619166.

References

1. Baumanns, S., Clemens, M., Schöps, S.: Structural aspects of regularized full Maxwell electrodynamic potential formulations using FIT. In: Proceedings of the 2013 International Symposium on Electromagnetic Theory, ISET, 24PM1C-01, pp. 1007–1010 (2014)
2. Schoenmaker, W., Chen, Q., Galy, P.: Computation of self-induced magnetic field effects including the Lorentz force for fast-transient phenomena in integrated-circuit devices. IEEE Trans. Comput. Aided Des. Integr. Circuits Syst. **33**, 893–902 (2014)

Chapter 4
Sensitivity of Lumped Parameters to Geometry Changes in Finite Element Models

Sebastian Schuhmacher, Carsten Potratz, Andreas Klaedtke, and Herbert De Gersem

Abstract The functional behavior of an electronic device is represented by an idealized circuit. Undesired parasitic interactions, such as electromagnetic-compatibility (EMC) problems, are modeled by additional lumped elements in the circuit. Device design parameters, e.g. partial inductances, must be optimized to improve EMC. This paper presents a sensitivity analysis method which relates changes to circuit parameters to changes to 3D model parameters.

4.1 Introduction

The increasing integration density in modern electromechanical systems requires considering electromagnetic-compatibility (EMC) issues at an early design stage in order to avoid costly changes later on. Historically, the electronic behavior of the system is modeled using a purely functional electronic circuit. Due to undesired parasitic effects, the physical realization behaves differently than the idealized model. In [1, 2], a method was proposed for automating the extraction of lumped elements between a given set of terminals from a 3D finite-element (FE) model. This approach is comparable to the more common model order reduction (MOR) [3] and partial element equivalent circuit (PEEC) [4, 5] techniques but has as a benefit that it preserves interpretability, as the reduced model still embeds the functional aspects of the circuit and merely adds additional lumped elements accounting for parasitics. This enables engineers to use their intuition in designing mitigation strategies.

S. Schuhmacher (✉) · A. Klaedtke · C. Potratz
Robert Bosch GmbH - Corporate Research, Renningen, Germany
e-mail: Sebastian.Schuhmacher@de.bosch.com; Andreas.Klaedtke@de.bosch.com

H. De Gersem
TU Darmstadt, Darmstadt, Germany
e-mail: degersem@temf.tu-darmstadt.de

© Springer International Publishing AG, part of Springer Nature 2018
U. Langer et al. (eds.), *Scientific Computing in Electrical Engineering*,
Mathematics in Industry 28, https://doi.org/10.1007/978-3-319-75538-0_4

The sensitivity of the EMC performance on the circuit parameters is easily calculated. Once this is known, the question arises how to adapt the physical realization model to improve the EMC. To guide this adaptation, we propose a method that visualizes all parasitic dependencies and thus allows to make informed decisions on geometry modifications.

Section 4.2 first recapitulates the circuit parameter extraction approach. This is followed by the derivation of the sensitivity analysis using an adjoint technique [6]. It will be discussed how changes in geometry or material parameters relate to changes in the extracted circuit parameters, and it is shown how sensitivity maps are generated. Academic and industrial example applications are presented in Sect. 4.3.

4.2   Extraction and Sensitivity Analysis of Circuit Parameters

A robust extraction of an equivalent electric circuit (EEC) is achieved if ohmic losses are extracted in the stationary-current approximation to Maxwell's equations, static capacitances in the electrostatic approximation [7] and inductances and coupling factors in Darwin's approximation [8]. This section briefly recapitulates these three extraction approaches and then derives the sensitivity analysis for each of them. The sensitivity analysis is achieved by computing the derivatives of the lumped element parameters in the EEC to all model geometry and material parameters. The results are presented in the form of sensitivity maps. The adjoint variable method [6] allows for an efficient computation of the derivative of an extracted circuit parameter with respect to all model parameters.

4.2.1   Partial Inductances

Darwin's approximation is used to extract partial inductances and equivalent partial capacitances as it describes capacitive as well as inductive behavior [9]. It provides an approximation to Maxwell's equations that excludes wave propagation phenomena and naturally confirms to a network description [7].

The formulation reads

$$\left(\mathbf{S} + s^2 \mathbf{T}\right) \mathbf{x} = \mathbf{y} \quad , \tag{4.1}$$

where

$$\mathbf{S} = \begin{pmatrix} \varepsilon \Delta & 0 \\ 0 & 0 \end{pmatrix}; \quad \mathbf{T} = \begin{pmatrix} -\varepsilon^2 \mu & \varepsilon \nabla \cdot \\ \varepsilon \nabla & -\nabla \times \mu^{-1} \nabla \times \end{pmatrix}; \quad \mathbf{x} = \begin{pmatrix} s^2 \varphi \\ \mathbf{E}_\sigma \end{pmatrix}; \quad \mathbf{y} = s^3 \begin{pmatrix} \varepsilon g \\ -\nabla \mu^{-1} g \end{pmatrix} \quad ,$$

ε is the permittivity, μ the permeability, $s = i\omega$ the Laplace variable, ω the angular frequency, φ the electric scalar potential, $\mathbf{E}_\sigma$ the electric field strength related to currents within the model and g an auxiliary field calculated a-priori by solving a magnetostatic problem for the external currents. A derivation and more detailed discussion of this formulation can be found in [10] and [2]. The formulation is discretized by the FE method. For conciseness, the discrete counterpart of (4.1) keeps the same notation. Hence, $\mathbf{x}$ now collects the degrees of freedom (DOFs) for $s^2\varphi$ and $\mathbf{E}_\sigma$.

Once the field solution $\mathbf{x}$ to (4.1) is known for as many orthogonal excitations as there are partial inductances (collected by the extended current matrix $\mathbf{I}$), the impedance matrix $\mathbf{Z}$ is calculated by:

$$\mathbf{Z}(s) = s^{-2}\,(\mathbf{Px})\,\mathbf{I}^{-1} \quad . \tag{4.2}$$

The projection operator $\mathbf{P}$ links the electrical scalar potential φ at the FE DOFs to the potentials at the vertices in the EEC. The network equivalent for the impedance matrix $\mathbf{Z}$ is given by a parallel connection of a capacitance and an inductance:

$$\mathbf{Z}(s) = ((s\mathbf{L})^{-1} + s\mathbf{C})^{-1} \quad . \tag{4.3}$$

By computing the impedance matrices at multiple frequencies well below the first resonance of the system, a least squares fit leads to the inductance.

In order to calculate the sensitivities of the inductances, we first have to calculate the change of the impedance $\mathbf{Z}$ with regards to a model parameter p_i:

$$\frac{\mathrm{d}\mathbf{Z}}{\mathrm{d}p_i} = \left(\frac{\partial \mathbf{Z}}{\partial \mathbf{x}}\right)^T \frac{\mathrm{d}\mathbf{x}}{\mathrm{d}p_i} \quad . \tag{4.4}$$

The second factor is the change of the solution vector $\mathbf{x}$ by the model parameter p_i and can be obtained using Eq. (4.1):

$$\frac{\mathrm{d}\mathbf{x}}{\mathrm{d}p_i} = \left(\mathbf{S} + s^2\mathbf{T}\right)^{-1} \left[\frac{\mathrm{d}\mathbf{y}}{\mathrm{d}p_i} - \left(\frac{\mathrm{d}\mathbf{S}}{\mathrm{d}p_i} + s^2\frac{\mathrm{d}\mathbf{T}}{\mathrm{d}p_i}\right)\mathbf{x}\right] \quad . \tag{4.5}$$

We use the adjoint technique, which requires the so-called adjoint solution λ which needs to be computed only once for each EEC parameter according to

$$\left(\mathbf{S} + s^2\mathbf{T}\right)^T \lambda = \frac{\partial \mathbf{Z}}{\partial \mathbf{x}} \quad . \tag{4.6}$$

The sensitivities are then

$$\frac{\mathrm{d}\mathbf{Z}}{\mathrm{d}p_i} = \lambda^T \left[\frac{\mathrm{d}\mathbf{y}}{\mathrm{d}p_i} - \left(\frac{\mathrm{d}\mathbf{S}}{\mathrm{d}p_i} + s^2\frac{\mathrm{d}\mathbf{T}}{\mathrm{d}p_i}\right)\mathbf{x}\right] . \tag{4.7}$$

The adjoint technique avoids the costly matrix inversion in Eq. (4.5). The matrices $\mathrm{d}\mathbf{S}/\mathrm{d}p_i$ and $\mathrm{d}\mathbf{T}/\mathrm{d}p_i$ have to be computed for every parameter but are very sparse and can thus be assembled efficiently. This overall method is therefore a fast and efficient way to compute the sensitivities of a few quantities with respect to a much larger number of 3D model parameters.

4.2.2   Capacitances and Conductances

The extraction of capacitances C and conductances G from the field model is accomplished by matching the electric energy $W_{EEC} = W_{3D}$ and the power loss $P_{EEC} = P_{3D}$ between circuit and field model:

$$\tfrac{1}{2}C\,U^2 = \tfrac{1}{2}\int (\varepsilon\,\nabla\varphi)\cdot(\nabla\varphi)\;\mathrm{d}V = \tfrac{1}{2}\mathbf{x}_\varepsilon{}^\mathrm{T}\mathbf{L}_\varepsilon\mathbf{x}_\varepsilon \quad; \tag{4.8}$$

$$G\,U^2 = \int (\sigma\,\nabla\varphi)\cdot(\nabla\varphi)\;\mathrm{d}V = \mathbf{x}_\sigma{}^\mathrm{T}\mathbf{L}_\sigma\mathbf{x}_\sigma \quad, \tag{4.9}$$

where U is the potential difference applied between the nodes in the EEC and accordingly on the inner boundaries of the field model. The discrete Laplacians $\mathbf{L}_\varepsilon$ and $\mathbf{L}_\sigma$ correspond to the electrostatic and stationary-current formulations respectively, and $\mathbf{x}_\varepsilon$ and $\mathbf{x}_\sigma$ are the solutions of the discrete Laplace problems accomplished with the above mentioned imprinted potential boundary conditions.

Applying the adjoint sensitivity method [6] to (4.8) or (4.9), the change of a circuit parameter $Q = (C, G)$ to a model parameter p reads:

$$\frac{\mathrm{d}Q}{\mathrm{d}p} = 2\mathbf{x}_{\varepsilon,\sigma}^\mathrm{T}\left(\frac{\mathrm{d}\mathbf{b}^{\varepsilon,\sigma}}{\mathrm{d}p} - \frac{\mathrm{d}\mathbf{L}^{\varepsilon,\sigma}}{\mathrm{d}p}\mathbf{x}_{\varepsilon,\sigma}\right) + \mathbf{x}_{\varepsilon,\sigma}^\mathrm{T}\frac{\mathrm{d}\mathbf{L}^{\varepsilon,\sigma}}{\mathrm{d}p}\mathbf{x}_{\varepsilon,\sigma} = \mathbf{x}_{\varepsilon,\sigma}^\mathrm{T}\frac{\mathrm{d}\mathbf{L}^{\varepsilon,\sigma}}{\mathrm{d}p}\mathbf{x}_{\varepsilon,\sigma} \quad, \tag{4.10}$$

where $\frac{\mathrm{d}\mathbf{b}^{\varepsilon,\sigma}}{\mathrm{d}p}$ denotes the change of the boundary condition terms with the imprinted potentials. In contrast to determining the sensitivities of the partial inductances, here, the adjoint solution does not have to be computed explicitly. As can be shown, the first term between the brackets in (4.10) does not contribute to the sensitivity, as the imprinted boundary potentials are independent from the geometric changes.

4.3   Application Examples

4.3.1   Plate Capacitor

An idealized plate capacitor with relative permittivity ε_r, surface area S and distance d between both plates, is considered. The analytic solutions for the capacitance and for its sensitivity to d are:

$$C_{\text{analytic}} = \varepsilon_0\varepsilon_r\frac{S}{d} \quad; \qquad \frac{\mathrm{d}C_{\text{analytic}}}{\mathrm{d}d} = -\varepsilon_0\varepsilon_r\frac{S}{d^2} \quad. \tag{4.11}$$

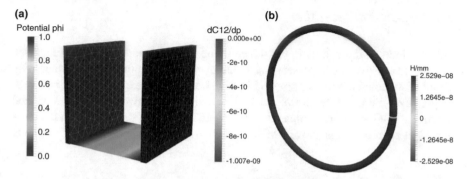

Fig. 4.1 Sensitivity maps generated by calculating the geometric sensitivity for predefined partitioned surface parts with respect to their normal vectors. The assigned sensitivities are represented by a color scale on the partitioned parts. (**a**) Sensitivity map for a plate capacitor, with surface area $S = 25\,\text{cm}^2$, plate distance $d = 4.2\,\text{cm}$ and relative dielectric permittivity between the plates $\epsilon_r = 80$. Also shown is the potential distribution between the plates. (**b**) Sensitivity map of a conductive ring with ring diameter $D = 200\,\text{mm}$ and wire diameter $d = 10\,\text{mm}$

The application of the described method to the 3D field model shown in Fig. 4.1a gives the results tabulated below. Magnetic boundaries were put around the capacitor, such that the electric field between the plates is perfectly perpendicular to the plates, which represents the "idealized" plate capacitor.

$C_{\text{calculated}}$	C_{analytic}	dC_{adjoint}/dd	dC_{analytic}/dd
42.16280 pF	42.16280 pF	$-10.06852\ \frac{\text{pF}}{\text{cm}}$	$-10.03876\ \frac{\text{pF}}{\text{cm}}$

4.3.2 Conducting Wire

To test the extraction method, a conducting ring with ring diameter D, wire diameter d and in a medium with permeability μ is considered (Fig. 4.1b). The analytic solutions for the self inductance L_{analytic}[11] and its derivative with respect to d are

$$L_{\text{analytic}} = \tfrac{1}{2}\mu D \left(\ln \frac{8D}{d} - 2 \right) \quad ; \quad \frac{dL_{\text{analytic}}}{dd} = -\mu \left(\frac{D}{2d} \right) \ . \tag{4.12}$$

The mesh faces of the ring (Fig. 4.1b) are displaced along their face normals. Note that the analytic solution is derived for a ring in free space whilst magnetic boundary conditions are used in the numerical experiment, with a distance to the object of four times the ring's diameter.

$L_{\text{calculated}}$	L_{analytic}	dL_{adjoint}/dd	dL_{analytic}/dd
384.4 nH	386.0 nH	$-25.29\ \frac{\text{nH}}{\text{mm}}$	$-25.13\ \frac{\text{nH}}{\text{mm}}$

4.3.3 Low-Pass π-Filter

The 3D model of the π-filter in Fig. 4.3 contains a coil that stands for the inductance and the two bails that represent the inductance of the capacitor. The manually created functional circuit is presented in Fig. 4.2b and the extracted EEC is shown in Fig. 4.2a. Figure 4.2c shows the network simulation results of the idealized functional low-pass filter compared to the results for the EEC. The EEC behavior is non-ideal at frequencies above 5 MHz. A sensitivity analysis on the network level

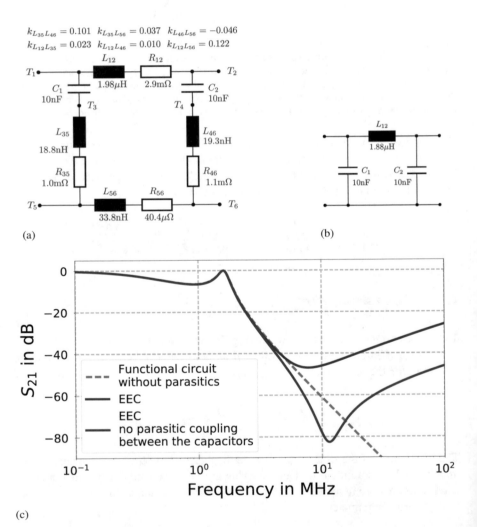

(a)

(b)

(c)

Fig. 4.2 (a) Extracted EEC (b) Idealized functional π-filter circuit (c) Filter transmission $S_{21}(f)$: Idealized circuit (b) (dashed green curve). EEC of (a) (red curve). EEC without coupling between the capacitors (blue curve)

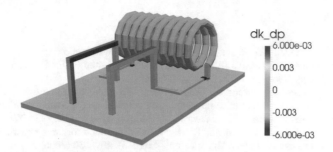

Fig. 4.3 Realization of a π-filter (only conducting parts are shown). Sensitivity map of the inductive coupling factor dk/dp_i between the modeled inductance of both capacitors with respect to the surface element normal vectors p_i. The values of the sensitivities are represented on the surface parts using a color scale

performed for the extracted EEC shows that the coupling between inductances L_1 and L_2 is responsible for most of the undesired high-frequency behavior. To verify this, the coupling factor k_{21} between the inductances L_1 and L_2 was set to 0. The corresponding result is compared to the raw EEC S-parameter in Fig. 4.2c. This change improves the filter performance by more than 20 dB.

Knowing that the coupling between the two bails improves the filter performance, its geometric sensitivity map is calculated by applying the adjoint sensitivity method. The geometric sensitivity map (Fig. 4.3) indicates that moving the surface of the bails to the outside and making the embedded area smaller decreases the coupling factor, whereas moving the surfaces to the inside increases the coupling, as expected.

4.4 Conclusions

In this paper, we provide an extension to a physically interpretable, reduced equivalent electric circuit extraction approach (as described in [1] and [2]). We provide the sensitivity analysis for passive lumped elements by using the adjoint technique. This method allows for an efficient computation of the derivatives of a lumped element parameter with regards to a large number of model parameters. The exemplary validation of the method presents interpretable sensitivity maps that show the sensitivity of a selected circuit parameter visualized on the geometry.

Acknowledgements The π-filter model used in Sect. 4.3.3 was kindly provided by Dr. Christoph Keller, Robert Bosch GmbH (CR/ARE1).

References

1. Traub, F., Hansen, J., Ackermann, W., Weiland, T.: Generation of physical equivalent circuits using 3d simulations. In: 2012 IEEE International Symposium on Electromagnetic Compatibility (EMC), pp. 486–491. IEEE, New York (2012)
2. Traub, F., Hansen, J., Ackermann, W., Weiland, T.: Automated construction of physical equivalent circuits for inductive components. In: International Symposium on Electromagnetic Compatibility (EMC Europe), Piscataway, NJ, 2013, pp. 67–72. IEEE, New York (2013)
3. Wittig, T., Schuhmann, R., Weiland, T.: Model order reduction for large systems in computational electromagnetics. Linear Algebra Appl. **415**(2–3), 499–530 (2006)
4. Heeb, H., Ruehli, A.: Three-dimensional interconnect analysis using partial element equivalent circuits. IEEE Trans. Circuits Syst. I: Fundam. Theory Appl. **39**(11), 974–982 (1992)
5. Bondarenko, N., Makharashvili, T., He, J., Berger, P., Drewniak, J., Ruehli, A.E., Beetner, D.G.: Development of simple physics-based circuit macromodel from PEEC. IEEE Trans. Electromagn. Compat. **58**(5), 1485–1493 (2016)
6. Belegundu, A.D.: Lagrangian approach to design sensitivity analysis. J. Eng. Mech. **111**(5), 680–695 (1985)
7. Raviart, P.-A., Sonnendrücker, E.: A hierarchy of approximate models for the Maxwell equations. Numer. Math. **73**(3), 329–372 (1996)
8. Larsson, J.: Electromagnetics from a quasistatic perspective. Am. J. Phys. **75**(3), 230–239 (2007)
9. Hansen, J., Potratz, C.: Capacity extraction in physical equivalent networks. In: 2015 IEEE International Symposium on Electromagnetic Compatibility (EMC), pp. 491–496. IEEE, New York (2015)
10. Traub, F., Hansen, J., Ackermann, W., Weiland, T.: Eigenmodes of electrical components and their relation to equivalent electrical circuits. In: 2013 IEEE International Symposium on Electromagnetic Compatibility (EMC), pp. 287–293. IEEE, New York (2013)
11. Paul, C.R.: Inductance: Loop and Partial. Wiley, Oxford (2010)

Chapter 5
Electro-Thermal Simulations with Skin-Layers and Contacts

Christoph Winkelmann, Raffael Casagrande, Ralf Hiptmair,
Philipp-Thomas Müller, Jörg Ostrowski, and Thomas Werder Schläpfer

Abstract We show a coupled electro-thermal simulation of a large, complex industrial device that yields a steady state temperature distribution with only small deviations from measurements. Firstly, the Ohmic losses in the conductors are calculated by a FEM-solver for the time-harmonic full Maxwell equations. To this end, we introduce a model to account for electric contact resistances, and a gradient based error indicator for adaptive mesh refinement. Secondly, the steady state temperature distribution is computed by a commercial CFD solver, taking into account convective and radiative cooling to balance the Ohmic heating. Theoretical arguments and simulation results hint that good predictions of total Ohmic losses and temperature distributions can be obtained on comparably coarse meshes which do not fully resolve the skin layer.

5.1 Background

Industrial power devices are usually large and geometrically complex. Examples of such devices are transformers or circuit breakers (CB). During nominal operation, the alternating current produces Ohmic losses that heat up the device. Losses that occur at the connections of the parts due to contact resistances sometimes amount up to 50% of all Ohmic losses. The devices are cooled by convection and by radiation.

C. Winkelmann (✉)
ABB Switzerland Ltd., Corporate Research, Baden-Dättwil, Switzerland

ETH Zürich, Seminar for Applied Mathematics, Zürich, Switzerland
e-mail: christoph.winkelmann@ch.abb.com

R. Casagrande · R. Hiptmair
ETH Zürich, Seminar for Applied Mathematics, Zürich, Switzerland

P.-T. Müller
RWTH Aachen, Aachen, Germany

J. Ostrowski · T. W. Schläpfer
ABB Switzerland Ltd., Corporate Research, Baden-Dättwil, Switzerland

© Springer International Publishing AG, part of Springer Nature 2018
U. Langer et al. (eds.), *Scientific Computing in Electrical Engineering*,
Mathematics in Industry 28, https://doi.org/10.1007/978-3-319-75538-0_5

In order to prevent damage, the temperature needs to be kept below device-specific limits everywhere. The experimental determination of the temperature distribution is possible but expensive. Simulations are a much cheaper and more enlightening alternative [1].

To do that, one first needs to calculate the Ohmic loss distribution in the device. Thereby, contact resistances that occur at the mentioned electrical connections cannot be neglected. Moreover, the skin layers are of particular concern as their thickness may be orders of magnitude smaller than the dimension of the device and therefore require—at first sight—a prohibitive number of mesh elements to resolve them.

In a former research project, we developed a time-harmonic $\mathbf{A} - \varphi$ based full Maxwell solver in Coulomb gauge that is stable in the low frequency limit [2]. We use this existing solver for the electromagnetic part of the coupled electro-thermal simulation. We develop models for the electrical contacts and for adaptive mesh refinement, and implement them there. They are introduced in Sects. 5.2 and 5.3, respectively. In Sect. 5.4, we analyze the convergence of the predictions of the Ohmic losses and the steady state temperature distribution under mesh refinement.

The steady state temperature distribution is calculated by using the commercial CFD solver ANSYS Fluent [3]. In Sect. 5.5 we show an electro-thermal simulation of a CB and compare the results with measurements.

5.2 Electric Contacts

Electric contact resistances (ECR) are a consequence of the roughness of the contacting surfaces [4]. As the actual thin layer with increased resistance and strong voltage drop cannot be resolved by the mesh, we introduce an actual jump in the voltage. We model this jump by adding to the initial formulation [2] a function s in the electric scalar potential $\varphi = \hat{\varphi} + s$ which is discontinuous at the surface Γ of the contact. Herein, $\hat{\varphi}$ represents the continuous part of the potential, see Fig. 5.1. The test function for the scalar potential is modified analogously as $\varphi' = \hat{\varphi}' + s'$. Since the total current is divergence-free, we find

$$0 = \int_{V_\Gamma^C} \mathrm{div} \mathbf{j}^t \, s' \, \mathrm{d}V \quad \Longrightarrow \quad 0 = \int_{\partial V_\Gamma^C} \mathbf{j}^t \cdot \mathbf{n} \, s' \, \mathrm{d}S - \int_{V_\Gamma^C} \mathbf{j}^t \, \mathbf{grad} \, s' \, \mathrm{d}V.$$

Herein, V_Γ^C is an adjacent volume inside the conductor on the side of the discontinuity where s and s' have support, shaded in blue in Fig. 5.1. The jump occurs only at the contact surface Γ. In the boundary integral, we neglect currents over the part of ∂V_Γ^C which coincides with the boundary of the conductor, and note that $s' = 0$ on the part inside the conductor away from the contact. Hence, only the part over Γ remains. There we assume that the electric field at the contact with very small

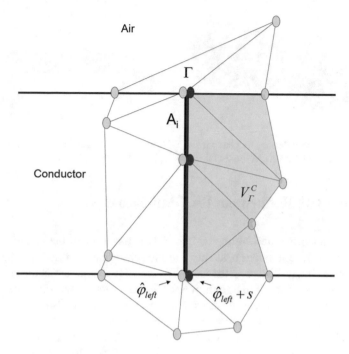

Fig. 5.1 FE approach for the electric contact

thickness d is given by $\mathbf{E} = \frac{s}{d}\, \mathbf{n}$. Then it follows

$$0 = \int_{\Gamma} (\sigma_{\Gamma} + i\omega\epsilon_{\Gamma})\,\frac{s}{d}\, s'\, \mathrm{d}S + \int_{V_{\Gamma}^{C}} (\sigma + i\omega\epsilon)(\mathbf{grad}(\hat{\varphi} + s) + i\omega\mathbf{A})\, \mathbf{grad}\, s'\, \mathrm{d}V.$$

Herein, σ is the electric conductivity, ϵ the permittivity, $\sigma_{\Gamma}, \epsilon_{\Gamma}$ the respective quantities inside the electrical contact, ω the angular frequency, and $\mathbf{A}$ the magnetic vector potential. Since we aim to compute electro-thermal phenomena at low frequencies, we neglect all displacement currents. Introducing the contact resistance $R_{\Gamma} = d/(|\Gamma|\sigma_{\Gamma})$, where $|\Gamma|$ denotes the area of Γ, we obtain the implemented formulation [5]:

$$0 = \frac{1}{R_{\Gamma}|\Gamma|} \int_{\Gamma} s\, s'\, \mathrm{d}S + \int_{V_{\Gamma}^{C}} \sigma (\mathbf{grad}(\hat{\varphi} + s) + i\omega\mathbf{A})\, \mathbf{grad}\, s'\, \mathrm{d}V.$$

We have tested the formulation with several configurations, see Fig. 5.2.

We plan to validate the formulation in the future by comparison of simulations and experiments of industrial power devices.

Fig. 5.2 Left: Current density in a bar with an ECR in the center at 50 Hz. Right: Current density in a setup with two ECRs $R_1/R_2 = 2$ at 0 Hz $\Rightarrow I_1/I_2 = 1/2$

5.3 Adaptive Refinement for Ohmic Losses

We perform adaptive mesh refinement to reduce the error in the calculated Ohmic losses. The mesh refinement is adaptive in the sense that we refine the mesh where we expect the biggest error in the Ohmic losses. Hence, an estimate or at least an indication of this error has to be computed. One could use a rigorous error estimator, e.g. as presented in [6–8]. Instead, we choose to develop an ad-hoc error indicator. We will refine all conductor cells with an error indicator greater than a threshold which is chosen such that the number of cells of the final mesh does not exceed a specified hard limit. Our error indicator η_T for cell T is defined as the maximum of a gradient indicator and a skin indicator: $\eta_T = \max(\eta_{T,g}, \eta_{T,s})$.

The gradient indicator $\eta_{T,g} = |T|\sqrt{q_T/\sigma_T}\,\max_{T' \in N(T)}\|\mathbf{j}_T - \mathbf{j}_{T'}\|_2$ approximates the error in the Ohmic losses in cell T using loss density q, conductivity σ, and current density $\mathbf{j}$ on neighboring conductor cells $T' \in N(T)$.

If there is only one cell across the conductor thickness, neighboring conductor cells will have very similar values, hence $\eta_{T,g}$ will be very small, although there could be a very fine skin layer and strong under-resolution of the loss distribution. In these cases, the error is strongly underestimated by $\eta_{T,g}$. This issue is overcome by the skin indicator $\eta_{T,s} = q_T |T| \left(1 - e^{-h_T/\delta_T}\right)$ which approximates the value that $\eta_{T,g}$ would take if the actual current density in T is assumed to decay to a fictitious neighboring conductor cell like in a flat skin layer, using diameter h_T, skin depth $\delta_T = \sqrt{2/(\mu_T \sigma_T \omega)}$ and permeability μ of cell T. Therefore, $\eta_{T,s}$ is consistent with $\eta_{T,g}$ in terms of unit and scaling, and taking the larger of the two ensures that skin layers are robustly detected also on coarse meshes.

We use the bar with ECR from Fig. 5.2 (left) as a test case. Figure 5.3 shows the error indicator per cell for a coarse mesh. One can see that it reliably detects cells at the ECR, next to edges and to surfaces.

Using the cell-wise error indicator η_T we can also construct an error indicator η_P for the total losses on some part P of the conductor: $\eta_P^2 = \sum_{T \subset P} \eta_T^2$. As we show in Sect. 5.4.2 by comparing to a solution on a much finer mesh, this error indicator for parts corresponds well to the actual errors.

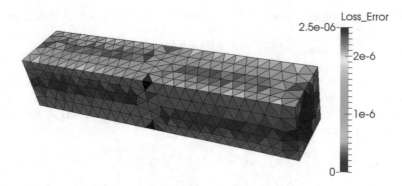

Fig. 5.3 Cell-wise error indicator η_T in Watt

5.4 Approximation Quality on Coarse Meshes

The Wiedemann-Franz law states that good electrical conductors are also good thermal conductors. Therefore, any non-uniform distribution of loss densities in a part made of a good electric conductor can be expected to be strongly smoothed out in the steady state temperature profile. The temperature can still differ significantly between parts, especially if they are separated by thermal contact resistances. Consequently, we can expect that the actual skin layer does not need to be resolved for the prediction of the steady state temperature distribution, as long as the total losses per part are well approximated.

5.4.1 Theory

A good approximation of the total losses per part can be expected from theory [9], as we will show below for a simplified setting. We consider the **A**-based variational formulation of the eddy-current problem in a simply connected domain Ω:

$$a(\mathbf{A}, \mathbf{A}') := \int_\Omega \mu^{-1}\mathbf{curl\,A} \cdot \mathbf{curl\,}\overline{\mathbf{A}'}\mathrm{d}V + i\omega \int_\Omega \sigma\mathbf{A} \cdot \overline{\mathbf{A}'}\mathrm{d}V = \int_\Omega \mathbf{j}_G \cdot \overline{\mathbf{A}'}\mathrm{d}V \quad (5.1)$$

with boundary condition $\mathbf{A} \times \mathbf{n} = \mathbf{0}$. Herein, $\mathbf{j}_G$ denotes the prescribed solenoidal generator current density. We solve the problem (5.1) on the quotient space $H = H_0(\mathbf{curl}, \Omega)/\{\mathbf{A} \in H_0(\mathbf{curl}, \Omega) : ||\mathbf{A}||_E = 0\}$, where $||\mathbf{A}||_E = |a(\mathbf{A}, \mathbf{A})|^{1/2}$ is the energy norm. Thus, the sesqui-linear form a satisfies the inf-sup condition. The total Ohmic losses on a part $P \subset \Omega$ is the continuously differentiable output functional $F(\mathbf{A}) := \omega^2 \int_P \sigma|\mathbf{A}|^2\mathrm{d}V$. We consider a Galerkin discretization of (5.1) on the space of first order edge elements R_h, for some mesh size h, and let $\mathbf{A}_h$ be a solution of it.

Because the output functional is differentiable, we can write the output error as

$$|F(\mathbf{A}_h) - F(\mathbf{A})| = |\langle F'(\mathbf{A}), \mathbf{A}_h - \mathbf{A}\rangle| + R(\mathbf{A}_h, \mathbf{A})$$

where the remainder R can be bounded as $|R(\mathbf{A}_h, \mathbf{A})| \leq C||\mathbf{A} - \mathbf{A}_h||_E^2$. By considering the dual problem in H: $a(\mathbf{A}', \mathbf{w}) = \langle F'(\mathbf{A}), \mathbf{A}'\rangle$, where $\mathbf{w} \in H$ is the dual solution, and using Galerkin orthogonality with an arbitrary $\mathbf{A}_h' \in R_h$, we can further estimate the output error as

$$|F(\mathbf{A}_h) - F(\mathbf{A})| = a(\mathbf{A}_h - \mathbf{A}, \mathbf{w} - \mathbf{A}_h') + R(\mathbf{A}_h, \mathbf{A})$$

$$\leq C_a||\mathbf{A}_h - \mathbf{A}||_E \inf_{\mathbf{A}_h' \in R_h} ||\mathbf{w} - \mathbf{A}_h'||_E + C||\mathbf{A} - \mathbf{A}_h||_E^2, \quad (5.2)$$

where C_a is the continuity constant of the sesqui-linear form a. While it is clear that $||\mathbf{A}_h - \mathbf{A}||_E \leq Ch$ if $\mathbf{A}$ is sufficiently smooth, the duality term $\inf_{\mathbf{A}_h' \in R_h} ||\mathbf{w} - \mathbf{A}_h'||_E$ requires further attention. Its behavior depends on the regularity of the dual solution $\mathbf{w}$ which in turn depends on the geometry of the conductor(s). If $\mathbf{w} \in H^s(\Omega)$ and **curl w** $\in H^s(\Omega)$ for $1/2 < s \leq 1$, then using local interpolation estimates we obtain $\inf_{\mathbf{A}_h' \in R_h} ||\mathbf{w} - \mathbf{A}_h'||_E \leq Ch^s$ for shape-regular sequences of meshes. Inserting this into eq. (5.2), we obtain $|F(\mathbf{A}_h) - F(\mathbf{A})| \leq Ch^{1+s}$ although we have only $||\mathbf{A} - \mathbf{A}_h||_E \leq Ch$ for first order edge elements.

In conclusion: While the local error of the current density $i\omega\sigma\mathbf{A}$ converges with first order in h, the error of the total losses per part converges with up to second order in h, provided the dual solution is sufficiently smooth.

5.4.2 Numerical Experiments

In order to further analyze the mesh quality required for electro-thermal simulations, we perform simulations on a series of meshes for the setup depicted in Fig. 5.4. A total current of 1250 A (peak) at 50 Hz is prescribed in the bar, which will induce an eddy current in the plate. Both parts are made of steel with a relative permeability of $\mu_R = 250$ (linear) and a conductivity of $\sigma = 5 \cdot 10^6$ S/m. The resulting skin depth δ is 2 mm. As reference values, we use values on a much finer mesh.

The relative errors in the Ohmic losses per part are plotted against the ratio of mesh cell size h (on the surface) over the skin depth δ in Fig. 5.5. It can be seen that an acceptable error of less than 4% in the bar can be reached when the cell size is twice as big as the skin depth, despite the fact that the Ohmic loss distribution shown in Fig. 5.4 is clearly a very bad approximation of reality. However, in the plate where the current is not prescribed but induced, the mesh needs to be about 3 times finer to reach the same level of accuracy. Note that on the two coarsest meshes, we have only one element in plate thickness which has to describe the current flowing in opposite directions on either side of the plate. We observe second order convergence

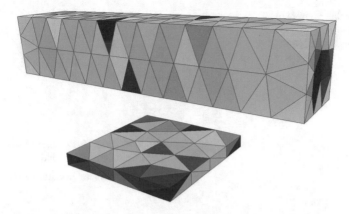

Fig. 5.4 Ohmic loss density on coarsest mesh of bar (10 by 10 by 50 mm, above) and plate (20 by 20 by 2 mm, below), 10 mm apart, two different color scales

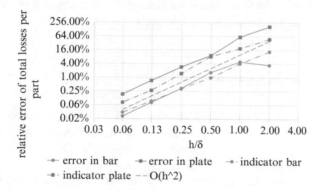

Fig. 5.5 Relative errors in Ohmic losses per part

of the errors predicted by theory in Sect. 5.4.1. Also, the error indicator per part η_P introduced in Sect. 5.3 predicts the actual error well.

In order to assess the required mesh quality for the loss computation of the coupled electro-thermal problem, we solve a stationary heat equation on each part with heat transfer boundary conditions on their boundary, using a heat transfer coefficient of $1000 \, \text{W}/(\text{m}^2\text{K})$. The relative errors in the maximum temperature rise on the surface of the respective parts are plotted against the ratio of mesh cell size h on the surface and skin depth δ in Fig. 5.6. By comparison to Fig. 5.5, it can be seen that the relative error of the temperature rise is essentially the same as the relative error of the Ohmic losses per part. We can conclude that even for iron, which is a relatively bad thermal conductor compared to usual materials like aluminum or copper, it is not necessary to fully resolve the actual current distribution. All that counts is the precision of the computation of the total losses, which confirms

Fig. 5.6 Relative errors in
maximum surface
temperature rise per part

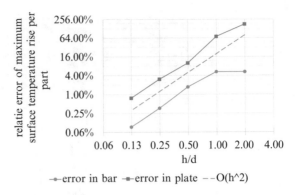

our initial expectation. However, if transient effects come into play rather than considering only the steady state, like e.g. for inductive hardening, resolution of the local distribution of the Ohmic losses is crucial. In these cases, it is recommended to consider adapted methods like the one presented in [10].

5.5 Electro-Thermal Simulation

If no electro-magnetic material parameter depends on temperature, a one-way coupling is exact: First, we perform a full Maxwell simulation to obtain the Ohmic losses. Then, we perform a simulation of convective and radiative cooling in ANSYS Fluent [3] with the Ohmic losses as source terms until a steady state is reached. Electric and thermal contact resistances are included, both of which are equally important. The mesh is different from the electric computation, and resolves thermal boundary layers. The Ohmic losses are interpolated from the electric to the thermal mesh.

We apply our simulation procedure to predict the steady temperature distribution in a CB at nominal operation. The CB is 7.5 m long, with wall thicknesses and skin depths in the order of 10 mm. The streamlines inside the CB are depicted in Fig. 5.7. They show the natural convection.

In Fig. 5.8 we compare the simulation with an experiment by plotting the mean temperature rise along both the inner conductor part and the enclosure. Simulation and experiment agree within 3 K, although there is often only one mesh cell in thickness direction. This again confirms that it is not necessary to fully resolve skin layers to obtain accurate predictions of steady state temperatures.

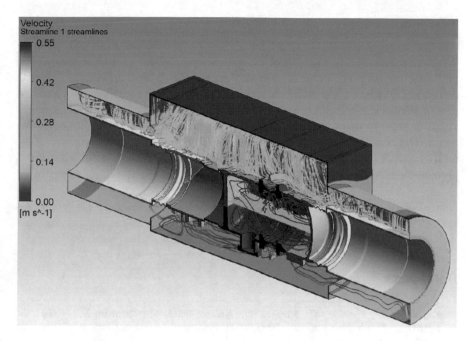

Fig. 5.7 Streamlines in CB

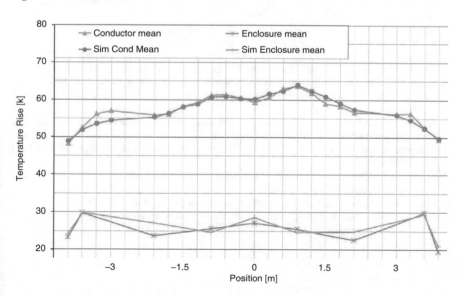

Fig. 5.8 Temperature rise along CB

5.6   Conclusions

We have shown that in order to obtain reliable predictions of temperature distributions in a power device, it is necessary to include electric and thermal contact resistances. However, it is sufficient to accurately predict the total Ohmic losses per part of the device, without necessarily resolving the skin layers, at least in *steady state*. In order to reach this goal, the required mesh resolution can in general be attained by moderate adaptive refinement, due to the quadratic convergence of the total losses per part. The error indicator, which is used for the refinement procedure, predicts the relative errors in the Ohmic losses accurately.

Acknowledgements This work has been co-funded by the Swiss Commission for Technology and Innovation (CTI).

References

1. Kaufmann, C., Günther, M., Klagges, D., Knorrenschild, M., Richwin, M., Schöps, S., ter Maten, J.: Efficient frequency-transient co-simulation of coupled heat-electromagnetic problems. J. Math. Ind. **4**, 1 (2014)
2. Hiptmair, R., Krämer, F., Ostrowski, J.: A robust Maxwell formulation for all frequencies. IEEE Trans. Magn. **44**(6), 682–685 (2008)
3. ANSYS Fluent. http://www.ansys.com/products/fluids/ansys-fluent
4. Holm, R.: Electric Contacts Handbook. Springer, New York (1958)
5. Mueller, P.T.: Macroscopic Electro Thermal Simulation of Contact Resistances. Bachelor's thesis, Rheinisch-Westfälische Technische Hochschule RWTH Aachen, Aachen (2016)
6. Beck, R., Hiptmair, R., Hoppe, R.H.W., Wohlmuth, B.: Residual based a posteriori error estimators for eddy current computation. ESAIM: Math. Model. Numer. Anal. **34**(1), 159–182 (2000)
7. Becker, R., Rannacher, R.: An optimal control approach to *a posteriori* error estimation in finite element methods. Acta Numer. **10**, 1–102 (2001)
8. Schöberl, J.: A posteriori error estimates for Maxwell equations. Math. Comput. **77**(262), 633–649 (2008)
9. Harbrecht, H.: On output functionals of boundary value problems on stochastic domains. Math. Methods Appl. Sci. **33**(1), 91–102 (2010)
10. Casagrande, R., Winkelmann, C., Hiptmair, R., Ostrowski, J.: A Trefftz method for the time-harmonic eddy current equation. In: Scientific Computing in Electrical Engineering SCEE 2016

Part II
Circuit and Device Modeling and Simulation

Circuit/device modeling and simulation are indispensable tools for the improvement of existing and the development of new electronic devices. Modeling and simulation help to reduce development costs and can substantially reduce the time to market for these devices. The increasing complexity of modern electronic devices poses significant new challenges for modeling and simulation with regard to the accuracy of the models and the efficiency of the computational methods for simulation. These aspects are addressed by the two contributions in this part, though with different focuses.

The first contribution is based on the keynote talk *"Gradient-Enhanced Polynomial Chaos Methods for Circuit Simulation"* by E.R. Keiter et al. A new approach to circuit level uncertainty quantification (UQ) by means of polynomial chaos expansion (PCE) methods is presented. PCE is a non-sampling, projection-based technique, in which parametric uncertainties are approximated using an expansion of orthogonal polynomials. This paper employs regression-based PCE, which requires less burdensome simulator modifications than fully intrusive Galerkin-based PCE. However, this comes at the cost of accuracy. The idea of enhancing the accuracy of regression-based PCE using gradient information is explored. The gradient information is provided by an intrusive adjoint sensitivity algorithm embedded in the circuit simulator.

The second paper, *"Coupled Circuit Device Simulation"* by K. Bittner et al., presents an approach for the coupling of electromagnetic field simulation with circuit simulation. A lumped device model is replaced by a full 3D model, which provides the data of the field model for use in the circuit simulator. The field model is based on (discretized) Maxwell equations provided by an interface to Magwel's electromagnetic simulation tool devEM. Numerical examples show that the coupled simulator can be used to test complex devices in the context of a larger circuit.

Chapter 6
Gradient-Enhanced Polynomial Chaos Methods for Circuit Simulation

Eric R. Keiter, Laura P. Swiler, and Ian Z. Wilcox

Abstract Uncertainty Quantification (UQ) is an important and emerging topic in electronic design automation (EDA), as parametric uncertainties are a significant concern for the design of integrated circuits. Historically, various sampling methods such as Monte Carlo (MC) and Latin Hypercube Sampling (LHS) have been employed, but these methods can be prohibitively expensive. Polynomial Chaos Expansion (PCE) methods are often proposed as an alternative to sampling. PCE methods have a number of variations, representing tradeoffs. Regression-based PCE methods, for example, can be applied to existing sample sets and don't require specific quadrature points. However, this comes at the cost of accuracy. In this paper we explore the idea of enhancing regression-based PCE methods using gradient information. The gradient information is provided by an intrusive adjoint sensitivity algorithm embedded in the circuit simulator.

6.1 Introduction

Sensitivity analysis and uncertainty quantification (UQ) are important capabilities for circuit simulation. In this paper, sensitivities refer to the derivatives of an objective function with respect to parameters. These parameter sensitivities give a local indication of the important parameters governing a response at a particular

E. R. Keiter (✉)
Electrical Models and Simulation, Sandia National Laboratories, Albuquerque, NM, USA
e-mail: erkeite@sandia.gov

L. P. Swiler
Optimization and Uncertainty Quantification, Sandia National Laboratories, Albuquerque, NM, USA
e-mail: lpswile@sandia.gov

I. Z. Wilcox
Component and Systems Analysis, Sandia National Laboratories, Albuquerque, NM, USA
e-mail: iwilcox@sandia.gov

© Springer International Publishing AG, part of Springer Nature 2018   55
U. Langer et al. (eds.), *Scientific Computing in Electrical Engineering*,
Mathematics in Industry 28, https://doi.org/10.1007/978-3-319-75538-0_6

point. UQ allows one to understand the probability distribution of the response, given probability distributions on the input parameters.

Sampling methods are commonly used to perform UQ. While sampling is an attractive approach for several reasons (e.g. it is repeatable given a particular seed, it is fault tolerant in the sense one can drop failed sample evaluations, and it is easy to understand), sampling suffers from the curse of dimensionality. A large number of samples are required to estimate the output statistics, especially to resolve small tail probabilities. The accuracy of the mean estimate obtained from a set of random samples exhibits $1/\sqrt{N}$ convergence, meaning that on average one needs to quadruple the number of sample points N to halve the error. Although many improvements on sampling schemes have been developed to overcome these limitations, such as Latin Hypercube Sampling [1, 2] and space-filling designs, the essential limitations of sampling still remain.

A recent interest in the computational simulation community is the use of more "embedded" UQ methods, in which the UQ algorithm is intrusively built into the simulator. As an example, in [3], an intrusive Galerkin based polynomial chaos expansion (PCE) method was demonstrated in a circuit simulator. However, the implementation required heavy instrumentation of the device models, which would be impractical in most production simulators.

There are categories of UQ method which require some simulator modification, but for which the required modifications are less burdensome than those necessary for fully intrusive Galerkin-based PCE. Specifically, if a simulator has been instrumented to efficiently produce parameter sensitivities [4–6], these can be used to enhance both the accuracy and runtime of several nominally non-intrusive UQ methods [7].

We outline the formulations for this UQ method, and demonstrate the computational savings that can be gained when using accurate sensitivities from an application code in the UQ process. The approaches and algorithms described in this paper are in implemented in two software frameworks: Xyce [8], a parallel circuit simulator developed at Sandia National Laboratories, and Dakota [9], an optimization and UQ toolkit also developed at Sandia. Both are open-source software packages available at https://info.sandia.gov/xyce and https://dakota.sandia.gov, respectively. However, it should be emphasized that the algorithms and approaches presented here are general, and applicable in other computational domains.

6.2 Transient Sensitivities

Many UQ techniques can be enhanced if the application code is able to produce parameter sensitivities with respect to objective functions of interest. In this paper, a high-level overview of direct and adjoint transient sensitivities are given. For a more detailed description the reader is encouraged to look at [4–6]. For this work, our interest is in transient dynamical systems represented by the differential-algebraic equation (DAE) form:

$$F(x, t, p) = \frac{dq(x(t, p), p)}{dt} + j(x(t, p), p) - b(t, p) = 0, \tag{6.1}$$

where $x \in \mathbb{R}^{n_x}$ is the DAE solution, which will satisfy $F = 0$ for all p. In circuit simulation, x consists of nodal voltages and branch currents. $p \in \mathbb{R}^{n_p}$ is a set of input parameters. q and j are functions representing the dynamic and static circuit elements respectively, and $b(t) \in \mathbb{R}^{n_x}$ is the input vector. In circuit analysis, q mostly contains capacitor charges, j contains resistance terms and b contains independent current and voltage sources. As such q, j and b are populated by the various circuit element models (also referred to as "compact models") supported by the circuit simulator. Transient analysis of Eq. (6.1) requires an implicit time integration method such as Backward Euler (BE) or the trapezoid rule. $F \in \mathbb{R}^{n_x}$ is the residual equation vector that is minimized by Newton's method at each time step to solve for x.

We are also interested in objective functions of the dynamical system, $O(x, p) \in \mathbb{R}^{n_O}$. For circuit simulation, the objective function could be a circuit output voltage, or something more complex, such as a signal delay. A sensitivity is the derivative of O with respect to p, which can be expressed using the chain rule giving:

$$\frac{dO}{dp} = -\frac{\partial O}{\partial x} \left(\frac{\partial F}{\partial x} \right)^{-1} \frac{\partial F}{\partial p} + \frac{\partial O}{\partial p}, \tag{6.2}$$

where x and F have the same meaning as in Eq. (6.1). The right-hand side of Eq. (6.2) contains the product of several matrices, which each have different dimensions. $\partial O / \partial x$ is of dimension $n_O \times n_x$. The Jacobian matrix $\partial F / \partial x$ is of dimension $n_x \times n_x$, and will generally be available in any simulator that solves Eq. (6.1) using implicit methods. The derivative vector $\partial F / \partial p$ is referred to as the "function derivative", is of dimension $n_x \times n_p$, and must be populated by the various compact device models. In modern circuit simulators, with complicated device compact models, computing $\partial F / \partial p$ can be challenging and may only be practical with automatic differentiation (AD). For this Xyce uses the Sacado AD library [10].

Sensitivities can be computed using two different methods; the direct method and the adjoint method. The difference between direct and adjoint is related to the order in which the terms of Eq. (6.2) are computed. For problems with large numbers of parameters n_p, and a small number of objectives n_O, the adjoint method is usually more efficient. For the opposite case, the direct method is a better choice. Transient direct and adjoint sensitivities are briefly described in Sects. 6.2.1 and 6.2.2 respectively.

6.2.1 Transient Direct Sensitivities

Transient direct sensitivities can be derived by following the approach described by
Hocevar [4]. For any integration method, a transient direct sensitivity DAE equation
can be derived by differentiating the original DAE (Eq. (6.1)) with respect to a
parameter, p:

$$\frac{dF(x, t, p)}{dp} = \frac{d}{dp}\left[\frac{dq(x(t), p)}{dt} + j(x(t), p) - b(t, p)\right] = 0 \qquad (6.3)$$

A numerical solution to Eq. (6.3) is obtained using an implicit time integration
method. If using BE, the expanded direct sensitivity DAE equation is determined
by substituting the BE formula for dq/dt and expanding the q and j derivatives
using the chain rule (for example $dq/dp = \partial q/\partial x \cdot \partial x/\partial p + \partial q/\partial p$). This gives:

$$\overbrace{\left[\frac{1}{h_i}\frac{\partial q_i}{\partial x_i} + \frac{\partial j_i}{\partial x_i}\right]}^{\text{Jacobian}}\frac{\partial x_i}{\partial p} = -\overbrace{\left(\frac{1}{h_i}\left[\frac{\partial q_i}{\partial p} - \frac{\partial q_{i-1}}{\partial p}\right] + \frac{\partial j_i}{\partial p} - \frac{\partial b_i}{\partial p}\right)}^{\text{Function Derivative}} + \overbrace{\frac{1}{h_i}\left[\frac{\partial q_{i-1}}{\partial x_{i-1}}\right]\frac{\partial x_{i-1}}{\partial p}}^{\text{Chain Rule term}},$$

11   (6.4)

where i is the time step index, and h_i is the time step size going from step $i-1$ to step
i. Similar formulas can be derived for other integration methods. Equation (6.4) is
solved at each time step once the Newton loop for the original DAE has converged.
The Jacobian matrix on the left-hand side of Eq. (6.4) is the same Jacobian as the one
used in the original DAE solve, so it can simply be reused. Note that the "function
derivative" on the right-hand side of Eq. (6.4) is equivalent to $\partial F/\partial p$ in Eq. (6.2),
and the Jacobian in Eq. (6.4) is the equivalent to $\partial F/\partial x$ from Eq. (6.2).

6.2.2 Transient Adjoint Sensitivities

Transient adjoint sensitivities [5, 6] can be broadly classified into two categories:
discrete adjoint sensitivities (in which one applies the adjoint operator after
discretizing the direct sensitivity DAE) and *continuous adjoint sensitivities* (in
which one applies the adjoint operator first, and then discretizes). For the sake of
brevity, this paper describes the discrete adjoint form [5].

For the discrete transient case, it is convenient to consider the entire transient in
block matrix form. If a transient simulation consists of N time points, then all the
time points can be considered in a single block matrix equation:

$$\mathbf{F} = \dot{\mathbf{Q}} + \mathbf{J} - \mathbf{B} = 0, \qquad (6.5)$$

where $\mathbf{F}$ is the block residual vector given by $\mathbf{F} = [F_0, F_1, \ldots, F_N]^T$. The other terms in the equation: $\mathbf{X}$, $\dot{\mathbf{Q}}$, $\mathbf{J}$, and $\mathbf{B}$ are block analogies of the original DAE equation terms: x, q, j, and b, respectively. For conventional time integration methods, the block Jacobian is a lower triangular block matrix:

$$\frac{\partial \mathbf{F}(\mathbf{X})}{\partial \mathbf{X}} = \begin{bmatrix} \left(\frac{\partial F_0}{\partial x_0}\right) & & & \\ \left(\frac{\partial F_1}{\partial x_0}\right) & \left(\frac{\partial F_1}{\partial x_1}\right) & & \\ \vdots & \vdots & \ddots & \\ & & & \left(\frac{\partial F_N}{\partial x_N}\right) \end{bmatrix}, \tag{6.6}$$

where the block linear system is:

$$\frac{\partial \mathbf{F}}{\partial \mathbf{X}} \Theta = \frac{\partial \mathbf{F}}{\partial p}, \tag{6.7}$$

and where $\Theta = [\Theta_0, \Theta_1, \ldots, \Theta_N]^T$ is the derivative of the solution $\mathbf{X} = [x_0, x_1, \ldots, x_N]^T$ with respect to a parameter value p. e.g., $\Theta_0 = dF_0/dp$. The block matrix is banded and lower triangular. Intuitively, solving this block linear system requires one to start with the upper left-hand corner of the matrix (at the first time point), and use forward substitution to solve the system. Doing this is analogous to integrating forward in time. For BE, the equivalent equation corresponding to block row i in Eq. (6.6) is:

$$\left(\frac{\partial F_i}{\partial x_i}\right) \frac{\partial x_i}{\partial p} = -\left(\frac{\partial F_i}{\partial x_{i-1}}\right) \frac{\partial x_{i-1}}{\partial p} + \frac{\partial F_i}{\partial p} \tag{6.8}$$

Equation (6.8) is equivalent to Eq. (6.4), when the residual F is expanded using the BE formula. $\partial F_i/\partial x_i$ is the Jacobian, $\partial F_i/\partial p$ the function derivative and $\partial F_i/\partial x_{i-1}$ the block matrix off-diagonal, or "chain rule term".

One can obtain the discrete adjoint form by taking the transpose of Eq. (6.6). The resulting block Jacobian has the form of an upper triangular matrix:

$$\left(\frac{\partial \mathbf{F}(\mathbf{X})}{\partial \mathbf{X}}\right)^T = \begin{bmatrix} \left(\frac{\partial F_0}{\partial x_0}\right)^T & \left(\frac{\partial F_1}{\partial x_0}\right)^T & & \\ & \left(\frac{\partial F_1}{\partial x_1}\right)^T & & \\ & & \ddots & \vdots \\ & & & \left(\frac{\partial F_N}{\partial x_N}\right)^T \end{bmatrix}, \tag{6.9}$$

where the block linear system is:

$$\left(\frac{\partial \mathbf{F}}{\partial \mathbf{X}}\right)^T \Theta_k^\star = \frac{\partial O_k}{\partial \mathbf{X}}, \tag{6.10}$$

and where $\Theta^\star$ is often referred to as the adjoint. There is a unique adjoint solution for each time point k. Similarly, the local objective function O at each time point k is considered to be a unique objective function, so O_k, is $\frac{\partial O_k}{\partial \mathbf{X}} = \left[0, 0, \ldots, \frac{\partial O_k}{\partial x_k}, \ldots, 0, 0\right]^T$. The matrix in Eq. (6.9) is upper triangular, so the solution requires a backsolve, starting in the lower right-hand corner at the final time point. This corresponds to integrating backward in time. As with direct methods, a variety of integration methods can be used to compute $\Theta^\star$. The BE form, corresponding to a single block row of the transposed block system, is given by:

$$\left[\frac{1}{h_i}\frac{\partial q_i}{\partial x_i} + \frac{\partial j_i}{\partial x_i}\right]^T \theta_i^\star = \left[\frac{1}{h_{i+1}}\frac{\partial q_{i+1}}{\partial x_{i+1}}\right]^T \theta_{i+1}^\star + \left(\frac{\partial O}{\partial x_i}\right)^T. \tag{6.11}$$

Equation (6.11) is evaluated in a loop stepping backward from the final time to the initial time.

Once $\Theta_k^\star$ has been computed for a specific time point k, it can be used to obtain dO_k/dp by taking the dot product with $\partial \mathbf{F}/\partial p$. In block matrix form this is given by:

$$\frac{dO_k}{dp} = \Theta_k^\star \cdot \frac{\partial \mathbf{F}}{\partial p}. \tag{6.12}$$

The derivative $\partial \mathbf{F}/\partial p$ is the function derivative. In the special case where $O = x$, then $\frac{\partial O_k}{\partial \mathbf{X}} = [0, 0, \ldots, 1, \ldots, 0, 0]^T$ and Eq. (6.12) provides dx/dp for a specific time point. If computing dx/dp for multiple time points, then a separate reverse integration and dot product evaluation is required for each one.

For transient adjoint sensitivities, it is necessary to completely solve the original DAE (Eq. (6.1)) for the entire time range first, before solving the adjoint equations to obtain sensitivities. Information must be saved during the forward solve in order to populate the Jacobians in Eq. (6.9), and the function derivatives in Eq. (6.12). For long transients this can require a lot of storage, a drawback of transient adjoints.

6.3 Polynomial Chaos Expansion Methods

Stochastic expansion UQ methods approximate the functional dependence of the simulation response on uncertain model parameters by expansion in a polynomial basis [11, 12]. The polynomials used are tailored to the characterization of the uncertain parameters. PCE is based on a multidimensional orthogonal polynomial approximation.

In PCE, the output response is modeled as a function of the input random variables using a carefully chosen set of polynomials. For example, PCE employs Hermite polynomials to model Gaussian random variables, as originally employed by Wiener [13]. Dakota implements the generalized PCE approach using the Wiener-Askey scheme [11], in which Hermite, Legendre, Laguerre, Jacobi, and generalized Laguerre orthogonal polynomials are used for modeling the effect of continuous random variables described by Gaussian, uniform, exponential, beta, and gamma probability distributions, respectively. These orthogonal polynomial selections are optimal for these distribution types since the inner product weighting function corresponds to the probability density functions for these continuous distributions.

To propagate input uncertainty through a model using PCE, Dakota performs the following steps: (1) input uncertainties are transformed to a set of uncorrelated random variables, (2) a basis such as Hermite polynomials is selected, and (3) the parameters of the functional approximation are determined. The general PCE for a response O has the form:

$$O(p) \approx \sum_{j=0}^{J} \alpha_j \Psi_j(p), \tag{6.13}$$

where each multivariate basis polynomial $\Psi_j(p)$ involves products of univariate polynomials that are tailored to the individual random variables. The response O is analogous to the objective function O described in Sect. 6.2, except that here the input parameters p are considered to be random variables and in Sect. 6.2 they are considered deterministic. If a total-order polynomial basis is used (e.g. a total order of 2 would involve terms whose exponents are less than or equal to 2, such as p_1^2, p_2^2, and $p_1 p_2$ but not $p_1^2 p_2^2$), the total number of terms N in a PCE of arbitrary order k for a response function involving n uncertain input variables is given by: $(n + k)!/(n!k!)$. If on the other hand, an isotropic tensor product expansion is used with order k in each dimension, the number of terms is $(k + 1)^n$. If the order k of the expansion captures the behavior of the true function, PCE methods will give very accurate results for the output statistics of the response.

In non-intrusive PCE, as in Dakota, simulations are used as black boxes and the calculation of the expansion coefficients α_j for response metrics of interest is based on a set of simulation response evaluations. To calculate these response PCE coefficients, two primary classes of approaches are used: spectral projection and regression. The spectral projection approach projects the response against each basis function $\Psi_j(p)$ using inner products and employs the polynomial orthogonality properties to extract each coefficient. Each inner product involves a multidimensional integral over the support range of the weighting function, which can be evaluated numerically using sampling, tensor-product quadrature, Smolyak sparse grid [14], or cubature [15] approaches. One advantage of PCE methods is their convergence rate [12]. For smooth functions (i.e., analytic, infinitely-differentiable) in L_2 (i.e., possessing finite variance), exponential convergence rates can be obtained under order refinement for integrated statistical quantities of interest such as mean and variance. A disadvantage of non-intrusive PCE methods is that they may not scale well to high dimensions. Recent research in adaptive refinement and sparse recovery methods strives to address this limitation [16].

In this work, we use regression-based PCE. Regression-based PCE approaches aim to solve the linear system:

$$\boldsymbol{\Psi}\boldsymbol{\alpha} \approx \boldsymbol{R} \qquad (6.14)$$

for a set of PCE coefficients $\boldsymbol{\alpha}$ that best reproduce a set of response values $\boldsymbol{R}$. The regression approach finds a set of PCE coefficients α_j which best match a set of response values obtained from a sampling study (e.g. a design of computer experiments producing an unstructured grid of sample points sometimes called collocation points.) on the density function of the uncertain parameters [17]. The convergence of regression-based PCE approaches has been studied. It is possible to bound the number of samples necessary to identify the coefficients in the PCE expansion by using the bounds on the spectral radius of a random matrix consisting of the sample points [18]. Convergence analyses focus on the number of samples and sampling approaches for stable and accurate solution recovery. The concept of a coherence parameter is used, which is a bound on the realized spectral radius of $\boldsymbol{W}\boldsymbol{\Psi}$, where $\boldsymbol{W}$ is a diagonal, positive definite matrix. Solution recovery of the PCE coefficients using regression PCE can be guaranteed with a number of samples that is proportional to the coherence times logarithmic factor in J, the total number of basis polynomials. In some cases, the number of samples required to recover the PCE coefficients scales linearly or nearly-linearly with the number of basis polynomials [18].

Additional regression equations can be obtained through the use of derivative information (gradients and Hessians) from each collocation point, which can aid in scaling with respect to the number of random variables, particularly for adjoint-based derivative approaches. This idea is the main subject investigated in this paper. The derivative equations are added to the set of regression equations as follows:

$$\frac{dO(p)}{dp} \approx \sum_{j=0}^{J} \alpha_j \frac{d\Psi_j(p)}{dp}. \qquad (6.15)$$

Equation (6.15) is simply the derivative of the PCE response equation (Eq. (6.13)) with respect to the random variables of the UQ analysis. The left-hand side is ideally provided by sensitivity calculations performed by the simulator, such as described in Sect. 6.2.

Various methods can be employed to solve Eq. (6.14). The relative accuracy of each method is problem-dependent. Traditionally, the most frequently used method has been least squares regression. However when $\boldsymbol{\Psi}$ is under-determined, minimizing the residual with respect to the L_2 norm typically produces poor solutions. Compressed sensing methods have been successfully used to address this limitation [19, 20]. Such methods attempt to only identify the elements of the coefficient vector α with the largest magnitude and enforce as many elements as possible to be zero. Such solutions are often called sparse solutions.

The convergence of gradient-enhanced regression PCE has been studied recently [21], where the authors show that the inclusion of derivative information and appropriate normalization will almost-surely lead to improved conditions for successful solution recovery. Reference [21] presents theoretical, probabilistic bounds regarding solution recovery for regression-based Hermite PCE with derivative information. This work suggests that adding gradients to the regression formulation will improve the solution recovery at a lower overall computational cost.

Dakota provides several algorithms that solve the regression formulations for PCE, including orthogonal matching pursuit, least angle regression (LARS), least absolute shrinkage (LASSO), basis pursuit, and a standard least squares. Typically, we recommend using least squares for over-determined systems and compressed sensing methods for under-determined systems, which is the case when the basis functions are augmented with additional basis functions representing gradient terms. Details of these methods are documented in the Linear Regression section of the Dakota Theory Manual [22].

6.4   Results for CMOS Inverter Circuit

In this section, we demonstrate the use of gradient-enhanced PCE methods on a five-stage CMOS inverter (Fig. 6.1). This circuit uses 10 instances of the BSIM6 [23] compact model, which in Xyce is instrumented with AD [10] to provide analytical parameter sensitivities (the "function derivative" term described in Sect. 6.2). The only other circuit element is a step input voltage source. The system to be solved is has 60 unknowns, most of which are nodal voltages. The dq/dx Jacobian is singular, so the system is a pure DAE system. The transistor models all include nonlinear capacitances. The capacitances from the first inverter form loops with the ideal voltage source input, meaning the circuit has a DAE index of two [24, 25].

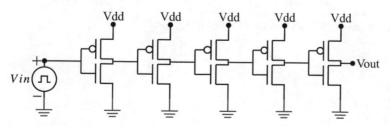

Fig. 6.1 CMOS circuit with five inverters

In digital circuits signal delay is an important performance metric. Capacitive effects are significant delay contributors, and in this circuit example each inverter stage adds to signal delay primarily through the gate oxide capacitors. Gate oxide thickness (referred to here as δ) is thus a critical uncertain parameter, and is specified as a parametric input to the BSIM6. For the purposes of this study, all five NMOS devices are assumed to have the same δ_N and all five PMOS devices are assumed to have the same δ_P, giving two uncertain scalar parameters. We model these as Gaussian-distributed uncertainties, centered around a nominal value with a standard deviation of 10% of nominal. The means of δ_N and δ_P were 1.74E−9m and 2.34E−9m, respectively. The other non-uncertain transistor parameters we used are taken from the BSIM6 benchmark tests.

The output of interest is the output voltage V_{out}, and a result from a forward Xyce calculation is plotted in Fig. 6.2. The left plot shows transient voltages for the input node, the third inverter output node, and the fifth inverter output node (V_{out}). In an ideal circuit, there would be no delay between the input and output transitions, but in this more realistic circuit that is not the case. The output voltages transition from the high state to low with some time delay after the step input, and each inverter adds additional delay to the signal. The V_{out} sensitivity with respect to δ_N and δ_P is shown on the right. Both are sharply peaked near the V_{out} transition. In this example $n_p = 2$, so the direct method (Sect. 6.2.1) was used to compute the sensitivities. However, an adjoint method (Sect. 6.2.2) produces identical results.

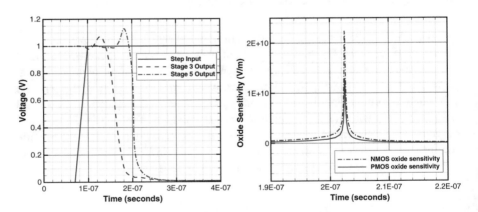

Fig. 6.2 Behavior of CMOS circuit exhibiting signal delay

To quantify delay, we used a generalized Elmore delay [6] as our objective function. If $g(t) = V_{out}$ is the transient response of a node in an electrical network to a step input, the delay T_D is approximated as the centroid of its time derivative $g'(t)$:

$$T_D = \frac{\int_A^B g'(t) \cdot t \cdot dt}{\int_A^B g'(t) dt} = \frac{\int_A^B g'(t) \cdot t \cdot dt}{g(B) - g(A)}. \tag{6.16}$$

The parameter derivative formula for T_D is given by:

$$\frac{dT_D}{dp} = \frac{\int_A^B \frac{dg'}{dp}(t) \cdot t \cdot dt - T_D \frac{d}{dp}[g(B) - g(A)]}{g(B) - g(A)}. \tag{6.17}$$

The quantities T_D, $dT_D/d\delta_N$ and $dT_D/d\delta_P$ are computed with Eqs. (6.16) and (6.17) using Xyce-computed values of V_{out}, $dV_{out}/d\delta_N$ and $dV_{out}/d\delta_P$ for a sequence of time steps. The integrals are approximated numerically using trapezoid rule. The time points A and B are simply the initial and final times of the simulation.

We performed UQ on the CMOS circuit using a variety of UQ techniques. As a baseline, we performed LHS with 100 and 1000 samples. Then, we performed PCE using a full tensor product quadrature of order 5 for each of the two input parameters, requiring 25 sample points. Finally, we performed two types of regression-based PCE. In the first, we used 30 samples without gradients. In the second, we used 10 samples, where each sample included two gradient values, $dT_D/d\delta_N$ and $dT_D/d\delta_P$. Thus, the last PCE calculation used 30 pieces of information comparable to the 30 sample regression PCE with no gradients, but only required 10 samples.

The use of sensitivities in performing uncertainty analysis is highlighted in Fig. 6.3 and Table 6.1. As shown in the figure, the cumulative distribution function (CDF), which gives the probability that T_D is less than a particular value, is almost the same for an LHS sample of size 1000 and all of the PCE methods. The CDF curves for LHS 1000 and for all of the PCE variants overlay each other. The only one that is noticeably different is the 100 sample LHS result. Table 6.1 shows that the mean T_D values are very similar, differing only in the fifth significant digit. Finally, the standard deviations show a little more variability, but again are reasonably close. We conclude that a PCE using sensitivities from Xyce (the 10 PCE regression case) performs comparably to 1000 LHS samples. Including gradients increases the cost per sample, but this additional cost is negligible for small problems. For small n_x, linear solves are less than 10% of total runtime. As a result, the two extra linear solves for each direct sensitivity time step do not incur much computational expense.

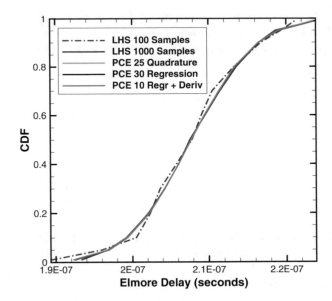

Fig. 6.3 CDFs for inverter delay (T_D)

Table 6.1 Various UQ method results

Number of samples and UQ method	T_D	
	Mean	Std dev.
100 LHS	2.0781E−7	6.6309E−9
1000 LHS	2.0782E−7	6.6935E−9
25 PCE quadrature	2.0783E−7	6.6954E−9
30 PCE regression	2.0783E−7	6.7131E−9
10 PCE regression with derivatives	2.0782E−7	6.7035E−9

6.5 Conclusions

This paper explored a new approach to circuit level UQ, based on gradient-enhanced PCE. PCE is a non-sampling, projection-based technique, in which parametric uncertainties are approximated using an expansion of orthogonal polynomials. Regression-based PCE can be enhanced by parametric sensitivities from the simulator, which offers the possibility of similar accuracy with fewer samples. In this paper, transient sensitivities are described, and the successful application of these sensitivities to gradient-enhanced PCE has been demonstrated.

Acknowledgements The authors gratefully acknowledge the anonymous reviewers for their careful reading of the manuscript and their valuable suggestions to improve various aspects of this paper.

This work was sponsored by the Laboratory Directed Research and Development (LDRD) Program at Sandia National Laboratories. Sandia National Laboratories is a multimission laboratory managed and operated by National Technology and Engineering Solutions of Sandia, LLC., a wholly owned subsidiary of Honeywell International, Inc., for the U.S. Department of Energy's National Nuclear Security Administration under contract DE-NA0003525.

References

1. McKay, M.D., Beckman, R.J., Conover, W.J.: A comparison of three methods for selecting values of input variables in the analysis of output from a computer code. Technometrics **21**(2), 239–245 (1979)
2. Stein, M.: Large sample properties of simulations using Latin hypercube sampling. Technometrics **29**(2), 143–151 (1987)
3. Strunz, K., Su, Q.: Stochastic formulation of SPICE-type electronic circuit simulation with polynomial chaos. ACM Trans. Model. Comput. Simul. **18**(4), 15:1–15:23 (2008)
4. Hocevar, D.E., Yang, P., Trick, T.N., Epler, B.D.: Transient sensitivity computation for MOSFET circuits. IEEE Trans. Comput. Aided Des. Integr. Circuits Syst. **CAD-4**(4), 609–620 (1985)
5. Gu, B., Gullapalli, K., Zhang, Y., Sundareswaran, S.: Faster statistical cell characterization using adjoint sensitivity analysis. In: Custom Integrated Circuits Conference, 2008, CICC 2008, pp. 229–232. IEEE, New York (2008)
6. Meir, A., Roychowdhury, J.: BLAST: Efficient computation of nonlinear delay sensitivities in electronic and biological networks using barycentric Lagrange enabled transient adjoint analysis. In: DAC'12: Proceedings of the 2012 Design Automation Conference, pp. 301–310. ACM, New York (2012)
7. Alekseev, A.K., Navon, I.M., Zelentsov, M.E.: The estimation of functional uncertainty using polynomial chaos and adjoint equations. Int. J. Numer. Methods Fluids **67**(3), 328–341 (2011)
8. Keiter, E.R., Aadithya, K.V., Mei, T., Russo, T.V., Schiek, R.L., Sholander, P.E., Thornquist, H.K., Verley, J.C.: Xyce parallel electronic simulator: users' guide, version 6.6. Technical Report SAND2016-11716, Sandia National Laboratories, Albuquerque, NM (2016)
9. Adams, B.M., Bauman, L.E., Bohnhoff, W.J., Dalbey, K.R., Eddy, J.P., Ebeida, M.S., Eldred, M.S., Hough, P.D., Hu, K.T., Jakeman, J.D., Rushdi, A., Swiler, L.P., Stephens, J.A., Vigil, D.M., Wildey, T.M.: Dakota, a multilevel parallel object-oriented framework for design optimization, parameter estimation, uncertainty quantification, and sensitivity analysis: version 6.2 users manual. Technical Report SAND2014-4633, Sandia National Laboratories, Albuquerque, NM (Updated May 2015). Available online from http://dakota.sandia.gov/documentation.html
10. Phipps, E.T., Gay, D.M.: Sacado Automatic Differentiation Package (2011). http://trilinos.sandia.gov/packages/sacado/
11. Xiu, D., Karniadakis, G.M.: The Wiener-Askey polynomial chaos for stochastic differential equations. SIAM J. Sci. Comput. **24**(2), 619–644 (2002)
12. Xiu, D.: Numerical Methods for Stochastic Computations: A Spectral Method Approach. Princeton University Press, Princeton (2010)
13. Wiener, N.: The homogeneous chaos. Am. J. Math. **60**, 897–936 (1938)
14. Smolyak, S.: Quadrature and interpolation formulas for tensor products of certain classes of functions. Dokl. Akad. Nauk SSSR **4**, 240–243 (1963)
15. Stroud, A.: Approximate Calculation of Multiple Integrals. Prentice Hall, Upper Saddle River (1971)

16. Constantine, P.G., Eldred, M.S., Phipps, E.T.: Sparse pseudospectral approximation method. Comput. Methods Appl. Mech. Eng. **229–232**, 1–12 (2012)
17. Walters, R.W.: Towards stochastic fluid mechanics via polynomial chaos. In: Proceedings of the 41st AIAA Aerospace Sciences Meeting and Exhibit, AIAA-2003-0413, Reno, NV (2003)
18. Hampton, J., Doostan, A.: Coherence motivated sampling and convergence analysis of least squares polynomial chaos regression. Comput. Methods Appl. Mech. Eng. **290**, 73–97 (2015)
19. Blatman, G., Sudret, B.: Adaptive sparse polynomial chaos expansion based on least angle regression. J. Comput. Phys. **230**(6), 2345–2367 (2011)
20. Doostan, A., Owhadi, H.: A non-adapted sparse approximation of PDEs with stochastic inputs. J. Comput. Phys. **230**(8), 3015–3034 (2011)
21. Peng, J., Hampton, J., Doostan, A.: On polynomial chaos expansion via gradient-enhanced l1-minimization. J. Comput. Phys. **310**, 440–458 (2016)
22. Adams, B.M., Bauman, L.E., Bohnhoff, W.J., Dalbey, K.R., Eddy, J.P., Ebeida, M.S., Eldred, M.S., Hough, P.D., Hu, K.T., Jakeman, J.D., Rushdi, A., Swiler, L.P., Stephens, J.A., Vigil, D.M., Wildey, T.M.: Dakota, a multilevel parallel object-oriented framework for design optimization, parameter estimation, uncertainty quantification, and sensitivity analysis: version 6.2 theory manual. Technical Report SAND2014-4253, Sandia National Laboratories, Albuquerque, NM (Updated May 2015). Available online from http://dakota.sandia.gov/documentation.html
23. Chauhan, Y.S., Venugopalan, S., Chalkiadaki, M.A., Karim, M.A.U., Agarwal, H., Khandelwal, S., Paydavosi, N., Duarte, J.P., Enz, C.C., Niknejad, A.M., Hu, C.: BSIM6: analog and RF compact model for bulk MOSFET. IEEE Trans. Electron Devices **61**(2), 234–244 (2014)
24. Günther, M., Feldmann, U.: The DAE-index in electric circuit simulation. Math. Comput. Simul. **39**(5–6), 573–582 (1995)
25. Bächle, S.: Index reduction for differential-algebraic equations in circuit simulation. Technical Report MATHEON 141, Technical University of Berlin, Germany (2004)

Chapter 7
Coupled Circuit Device Simulation

Kai Bittner, Hans Georg Brachtendorf, Wim Schoenmaker,
Christian Strohm, and Caren Tischendorf

Abstract The goal of coupled circuit-field simulation is to test a complex device described by a full 3D field model in the environment of a larger circuit. We present here an approach, which treats the field model as device with a large number of internal unknowns and equations.

7.1 Introduction

Today's most common lumped device models increasingly tend to loose their validity in circuit simulation due to the rapid technological developments, miniaturization and higher complexity of integrated circuits. This has motivated the idea of combining circuit simulation directly with distributed device models based on electromagnetic field equations to refine critical circuit parts (see e.g. [1–3]).

Our approach is a holistic simulation, where critical devices are described by a full 3D electro-magnetic field model given by Maxwell's equations. This field model is provided by the EM simulation tool devEM of Magwel [4, 5], which permits the Computer Aided Design of devices consisting of insulators, conductors, and semiconductors. Furthermore, devEM provides the field equations discretized by the finite integration method and offers DC, AC, and transient solvers as well as post processing and graphical output of the results.

Our approach for a coupled circuit/device simulation is to integrate the field equations into the circuit simulator LinzFrame of the University of Applied Sciences

K. Bittner (✉) · H. G. Brachtendorf
University of Applied Sciences of Upper Austria, Hagenberg, Austria
e-mail: Kai.Bittner@fh-hagenberg.at; Hans-Georg.Brachtendorf@fh-hagenberg.at

W. Schoenmaker
MAGWEL NV, Leuven, Belgium
e-mail: Wim.Schoenmaker@magwel.com

C. Strohm · C. Tischendorf
Dept. of Mathematics, Humboldt-University of Berlin, Berlin, Germany
e-mail: strohmch@math.hu-berlin.de; tischendorf@math.hu-berlin.de

© Springer International Publishing AG, part of Springer Nature 2018
U. Langer et al. (eds.), *Scientific Computing in Electrical Engineering*,
Mathematics in Industry 28, https://doi.org/10.1007/978-3-319-75538-0_7

of Upper Austria. To achieve this goal new device classes have been implemented. These device classes process data from the field equations, which are provided by two newly developed interfaces to the Magwel tool.

An overview of our coupling approach is presented in Sect. 7.2. In Sect. 7.3 the implementation of the coupling is described. Numerical examples are presented in Sect. 7.4

7.2 Coupled Simulation

The circuit is described as a network of devices, connected at nodes. We consider circuit equations in the charge/flux oriented modified nodal analysis (MNA) formulation, which yields a mathematical model in the form of a system of differential-algebraic equations (DAEs):

$$\frac{d}{dt}q\big(x(t)\big) + \underbrace{g\big(x(t)\big) + s(t)}_{f(x(t),t)} = 0. \tag{7.1}$$

The vector $x \in \mathbb{R}^n$ consists of all node potentials except ground as well as currents through inductors and voltage sources. The equations are obtained from Kirchhoff's current law for each node and additional equations for voltage sources and inductors. Kirchhoff's current law requires the sum off all currents into/from a node to be zero. These currents come from device terminals connected to the node. Except for inductors and voltage sources, where the currents are in the unknown vector, the terminal currents are functions of branch voltages, which in turn are the differences of node potentials.

For the coupling of electromagnetic field simulation with circuit simulation we replace lumped device models by a full 3D model based on (discretized) Maxwell equations. The discretized field model is provided by an interface to the electro magnetic simulation tool devEM of Magwel [4, 5]. The circuit equations (7.1) are complemented by the discretized field equations (as internal device equations) of the form

$$\frac{d}{dt}a\big(x(t)\big) + f\big(x(t)\big) = 0, \tag{7.2}$$

where the vector $x(t)$ of unknowns is extended by additional variables. The new system has now the same form as the original circuit equations (7.1) such that solvers for circuit simulation can be applied.

7.3 Implementation

The coupling was implemented in the circuit simulator LinzFrame from University of Applied Sciences of Upper Austria (see Fig. 7.1). The simulator core comprises the Modified Nodal Analysis (MNA) and an automatic differentiation suite [6],

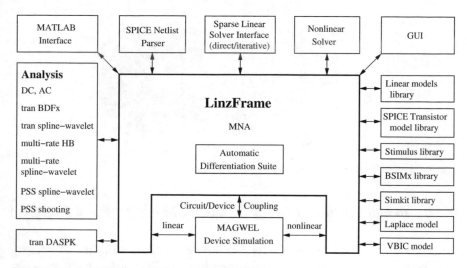

Fig. 7.1 The modular circuit-device simulator LinzFrame

model libraries to industry standard semiconductor models, and a source library including modulated sources (FSK, OFDM etc.). The analysis tools comprise standard simulation techniques such as AC, DC, TRAN, the latter with algebraic and trigonometric BDF techniques and dedicated techniques for RF circuits including HB, multi-rate spline-wavelet methods as well as shooting techniques. To deal with DAEs of a higher index LinzFrame is coupled to the DASPK initial value solver [7]. Numerical tools encompass several damped Newton techniques for solving nonlinear algebraic equations and an interface to several linear solvers, both direct and iterative Krylov subspace.

7.3.1 Lumped Device Models

As in many other circuit simulators an evaluation of the circuit equations is done by the evaluation of all devices in the circuit. Each device model is derived from the abstract C++ class Device. A device has K terminals, which connect to the circuit nodes $n_1, \ldots, n_K$ (cf. the example in Fig. 7.2 (left side) with $K = 4$). The device class has a method Device::eval(), which computes terminal currents and charges based on the node potentials of the terminal nodes and possible other internal unknowns. Such internal unknowns are currents through inductors and voltage sources as well as potentials of internal nodes. The terminal currents and charges are then added to Kirchhoff's current law of the corresponding nodes, yielding the corresponding entries of $q(x)$ and $f(x, t)$ in (7.1). Residuals for internal equations (inductors, voltage sources, internal nodes) are calculated and written as well. Furthermore, all partial derivatives are computed and written to the Jacobian matrix for Newtons methods.

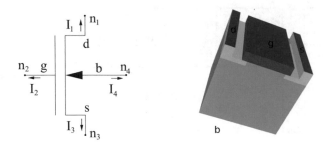

Fig. 7.2 Lumped model and 3D field model of a MOSFET

7.3.2 Replacement of a Lumped Device Model by a Field Model

To replace a lumped model by a 3D field model we have to include the (discretized) field equations. Furthermore, terminals need to be defined and terminal currents and charges have to be provided. To get access to the field equations an interface to the field solver **devEM** of Magwel was implemented to provide the necessary information. For the coupling several contact surfaces are defined in the field model, which act as device terminals in the circuit simulator as follows. The node potentials of the device terminals are used as boundary conditions for the electrical potential at the contact surface, while the terminal currents and charges for Kirchhoff's current law are obtained by (numerically) integrating current density and displacement over the contact surfaces (see e.g. Fig. 7.2 right side, where the contact surface are depicted in red). We have implemented a new abstract device class in LinzFrame:

```
class MagwelDevice : public Device
{
                              ⋮
```

This new device class provides an interface for the circuit simulator to handle the data from the Magwel interface. It defines the terminals as well as the additional unknowns and equation from the field model (see [8] for details).

There are two kinds of interfaces to the field solver. First there is a linear interface, which can be used if the device consists only of conductors and insulators so that all field equations are linear. Then, the discretized equations can be fully described by constant matrices. These matrices are provided by the field simulator in a pre-processing step. For implementation we have introduced the device class

```
class MagwelLinDev : public MagwelDevice
{
                              ⋮
```

Here, the matrices are computed in advance, written to a file, and then read by the constructor in the initialization. During the device evaluation these data are used to provide terminal currents and charges as well as the residuals for internal equations together with their partial derivatives.

If semiconductors are present in the device this approach does not work anymore since the equations become nonlinear. For this case a nonlinear interface is used (suitable for any device). Here the field solver provides for any vector of unknowns x the vectors $a(x)$ and $f(x)$ together with their Jacobians. Here we use the device class

```
class MagwelNLDev : public MagwelDevice
{
```

⋮

The constructor handles during the initialization only structural data (e.g. number of terminals, sparsity structure of the Jacobians). Residuals and matrices are then loaded during the device evaluation in each Newton step. Here the circuit simulator acts as master and the field simulator as slave.

Although the nonlinear interface can be used for any device it is recommended to use the linear interface if no semiconductors are present in the device. Using the precomputed data speeds the evaluation up essentially, compared to the repeated evaluation by the nonlinear interface.

7.4   Numerical Results

Figure 7.3 depicts a Colpitts oscillator where the lumped inductor is replaced by an on-chip element as well as the result of a transient simulation. The on-chip inductor is simulated by a full 3D electro magnetic field model using the linear interface.

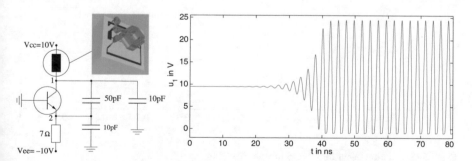

Fig. 7.3 Colpitts oscillator circuit with full 3D field models for inductors

The nonlinear interface was tested on the CMOS inverter in Fig. 7.4, where the MOSFET's have been simulated by a 3D field model. Due to the semiconductor materials the nonlinear interface was used here (Fig. 7.5).

Finally we show in Fig. 7.6 the results of a coupled multirate simulation (see [9–12] for details). The simulation was performed on a differential oscillator (Fig. 7.7), where again an electromagnetic field model was used for the inductors.

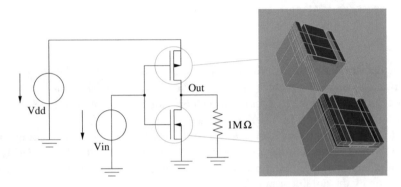

Fig. 7.4 CMOS inverter with full 3D field models for NMOS and PMOS

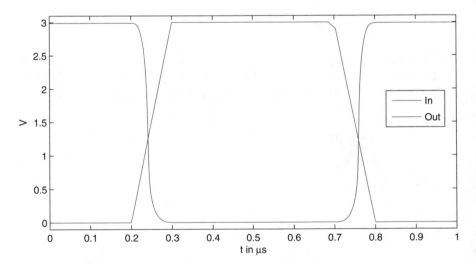

Fig. 7.5 Simulation result for CMOS inverter from Fig. 7.4

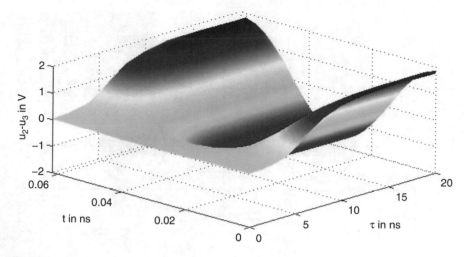

Fig. 7.6 Simulation result for differential oscillator from Fig. 7.7

7.5   Conclusions

The field simulator devEM of Magwel and the circuit simulator LinzFrame of
University of Applied Sciences of Upper Austria have been coupled for a cir-
cuit/device simulation. A lumped device model is replaced by a full 3D field model
providing the data of the field model for the use in the circuit simulator. Numerical
examples show that the coupled simulator can be used to test complex devices in
the environment of a larger circuit.

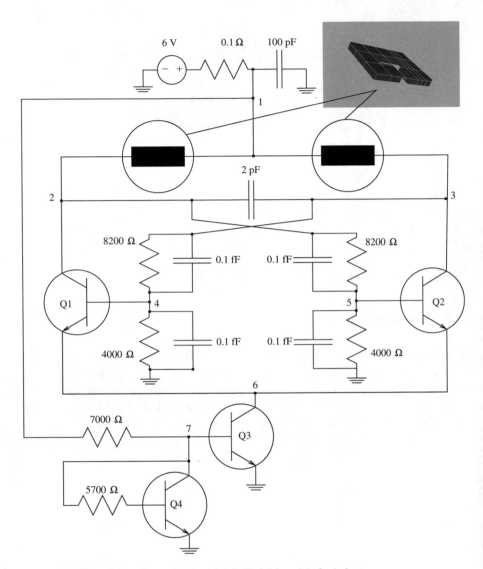

Fig. 7.7 Differential oscillator circuit with full 3D field models for inductors

Acknowledgements This work is funded by the European fp7 project nanoCOPS under grant 619166.

References

1. Engl, W.L., Laur, R., Dirks, H.K.: MEDUSA — a simulator for modular circuits. IEEE Trans. Comput. Aided Des. Integr. Circuits Syst. **1**(2), 85–93 (1982)
2. Mayaram, K., Pederson, D.O.: Coupling algorithms for mixed-level circuit and device simulation. IEEE Trans. Comput. Aided Des. Integr. Circuits Syst. **11**(8), 1003–1012 (1992)
3. Rotella, F.M.: Mixed circuit and device simulation for analysis, design, and optimization of opto-electronic, radio frequency, and high speed semiconductor devices. Ph.D. thesis, Stanford University (2000)
4. Schoenmaker, W., Chen, Q., Galy, P.: Computation of self-induced magnetic field effects including the lorentz force for fast-transient phenomena in integrated-circuit devices. IEEE Trans. Comput. Aided Des. Integr. Circuits Syst. **33**(6), 893–902 (2014)
5. Schoenmaker, W., Matthes, M., De Smedt, B., Baumanns, S., Tischendorf, C., Janssen, R.: Large signal simulation of integrated inductors on semi conducting substrates. In: 2012 Design, Automation & Test in Europe Conference & Exhibition, DATE, pp. 1221–1226. IEEE, New York (2012)
6. Feldmann, P., Melville, R.C., Long, D.E.: Efficient frequency domain analysis of large nonlinear analog circuits. In: Proceedings of the Custom Integrated Circuits Conference (1995)
7. Brenan, K.E., Campbell, S.L., Petzold, L.R.: Numerical Solution of Initial-Value Problems in Differential-Algebraic Equations. SIAM, Philadelphia (1996)
8. Schoenmaker, W., Meuris, P., Strohm, C., Tischendorf, C.: Holistic coupled field and circuit simulation. In: 2016 Design, Automation & Test in Europe Conference & Exhibition, DATE, pp. 307–312. IEEE, New York (2016)
9. Brachtendorf, H.G.: Theorie und Analyse von autonomen und quasiperiodisch angeregten elektrischen Netzwerken. Eine algorithmisch orientierte Betrachtung. Universität Bremen, Habilitationsschrift (2001)
10. Brachtendorf, H.G., Bittner, K., Laur, R.: Simulation of the steady state of oscillators in the time domain. In: 12th Design, Automation & Test in Europe (2012)
11. Bittner, K., Brachtendorf, H.G.: Optimal frequency sweep method in multi-rate circuit simulation. COMPEL **33**(4), 1189–1197 (2014)
12. Bittner, K., Brachtendorf, H.G.: Adaptive multi-rate wavelet method for circuit simulation. Radioengineering **23**(1), 300–307 (2014)

Part III
Coupled Problems and Multi-Scale Approaches in Space and Time

This part of the proceedings consists of three contributions on different aspects of coupled problems and their numerical treatment.

In the paper by K. Gausling and A. Bartel on *"Density Estimation Techniques in Cosimulation using Spectral- and Kernel Methods"*, the cosimulation of coupled systems of differential-algebraic equations is studied. In particular, the paper explores the influence of uncertainties on the contraction factor of a cosimulation scheme is explored. Two methods for estimating the probability density function (PDF) of the contraction factor are described. It introduces a new splitting approach for a benchmark problem of a coupled field/circuit model, and compares the two methods for estimating the PDF.

The focus of the second paper, *"Multirate DAE/ODE–Simulation and Model Order Reduction for Coupled Field-Circuit Systems"* by C. Hachtel et al., is on coupled field-circuit problems consisting of a slowly changing subsystem of magneto-quasistatic equations and a rapidly changing system of the surrounding circuit. These coupled problems lead to high-dimensional systems of partial-differential-algebraic equations and exhibit largely varying time scales. To exploit the different dynamical behaviors of circuit and field equations, multirate time integration schemes are presented. These schemes are then combined with a model order reduction technique for the slowly changing subsystem of magneto-quasistatic equations, which significantly decreases the computational effort required.

The third contribution, *"Modeling and Simulation of Electrically Controlled Droplet Dynamics"* by Yun Ouédraogo et al., addresses the droplet motion in strong electric fields described by a coupled system of a multiphase hydrodynamic problem and an electro-quasistatic problem. The hydrodynamic subsystem is modeled using the incompressible Navier-Stokes equations, where the phase boundaries are represented by a surface tension force. The electro-quasistatic subsystem is modeled by an irrotational electric field and the charge conversation equation. The model is discretized with the aid of a Finite Volume Method (FVM) on a fixed grid. As an application the simulation of controlled droplet generation in an electrically driven droplet generator is presented and compared with experimental data.

Chapter 8
Density Estimation Techniques in Cosimulation Using Spectral- and Kernel Methods

Kai Gausling and Andreas Bartel

Abstract When Co-simulation is applied to coupled differential algebraic equations, convergence can only be guaranteed if certain properties are fulfilled. However, introducing uncertainties in this mode may have great impact on these contraction properties and may destroy convergence. Hence one is interested to analyze the stochastic behavior of those properties. Within this paper we compare the Kernel Density Estimation technique and the spectral approach based on polynomial chaos expansion to measure the density of the contraction factor which may occur for coupled systems. Using the new R-splitting approach in a field/circuit coupled problem as benchmark, we clarify the benefits of both schemes.

8.1 Introduction

For networks which can be described by differential algebraic equations (DAEs) of index-1, the system can be described as differential algebraic initial-value problem

$$0 = \mathbf{g}(\mathbf{y}, \mathbf{z}), \quad \dot{\mathbf{y}} = \mathbf{f}(\mathbf{y}, \mathbf{z}) \quad \text{with } \mathbf{y}(t_0) = \mathbf{y_0} \tag{8.1}$$

with differential and algebraic unknown $\mathbf{y} \in [t_0, t_e] \to \mathbb{R}^{n_y}$, $\mathbf{z} \in [t_0, t_e] \to \mathbb{R}^{n_z}$.

Now co-simulation tries to compute the solution on $\Gamma = [t_0, t_e]$ iteratively by solving the decoupled subsystems. To this end synchronization points T_i are introduced, thus that $t_0 = T_0 < T_1 < \cdots < T_m = t_e$, with time window size $H_n := [T_{n+1} - T_n]$. Such a co-simulation scheme can be encoded by splitting functions $\mathbf{F}, \mathbf{G}$:

$$0 = \mathbf{G}\left(\tilde{\mathbf{y}}^{(k)}, \tilde{\mathbf{z}}^{(k)}, \tilde{\mathbf{y}}^{(k-1)}, \tilde{\mathbf{z}}^{(k-1)}\right), \quad \dot{\tilde{\mathbf{y}}} = \mathbf{F}\left(\tilde{\mathbf{y}}^{(k)}, \tilde{\mathbf{z}}^{(k)}, \tilde{\mathbf{y}}^{(k-1)}, \tilde{\mathbf{z}}^{(k-1)}\right) \tag{8.2}$$

K. Gausling (✉) · A. Bartel
Bergische Universität Wuppertal, Applied Mathematics and Numerical Analysis, Wuppertal, Germany
e-mail: gausling@math.uni-wuppertal.de; bartel@math.uni-wuppertal.de

© Springer International Publishing AG, part of Springer Nature 2018
U. Langer et al. (eds.), *Scientific Computing in Electrical Engineering*,
Mathematics in Industry 28, https://doi.org/10.1007/978-3-319-75538-0_8

The current iteration on H_n is denoted by (k), the old iterates are $(k-1)$. We drop the tilde in the following. Due to the decoupling a contraction factor

$$\alpha_n := \|\mathbf{G}_{\mathbf{z}^{(k)}}^{-1} \mathbf{G}_{\mathbf{z}^{(k-1)}}\|_{2,\infty}, \quad \text{with } \mathbf{G}_{\mathbf{z}^{(k)}} := \frac{\partial \mathbf{G}}{\partial \mathbf{z}^{(k)}}, \ \mathbf{G}_{\mathbf{z}^{(k-1)}} := \frac{\partial \mathbf{G}}{\partial \mathbf{z}^{(k-1)}}, \quad (8.3)$$

occurs, where convergence is only guaranteed if $\alpha_n < 1$ holds for $H_n < H_{\max}$, see [1, 2]. Consequently, co-simulation applied to coupled ordinary differential equations (ODEs) always convergences, see [3], and convergence of co-simulation applied to DAEs is only ensured for $\mathbf{G}_{\mathbf{z}^{(k)}}$, $\mathbf{G}_{\mathbf{z}^{(k-1)}}$ well structured, see [1]. In general the contraction factor α_n may depend on components from the model. Therefore, introducing uncertainties into the components may change the contraction properties, that is, (8.3) becomes stochastic with unknown probability density function (PDF). Here we discuss two different approaches namely the Kernel Density Estimation (KDE) technique and the spectral approach based on Polynomial Chaos (PC) expansion to estimate the underlying PDF of the contraction factor given in (8.3).

The paper is structured as follows: Chap. 2 introduces the lower bound estimation technique to measure the contraction factor. Chapter 3 gives inside into the KDE and spectral method where the sample of the contraction factor is used to estimate the underlying PDF. Chapter 4 is split into two parts. The first part introduces the new technique of R-splitting, where the contraction factor is defined by the ratio of resistances. This gives us the opportunity to measure the accuracy of the estimation introduced in Chap. 2. The second part compares the KDE and spectral method with respect to the attainable accuracy and its computational costs.

8.2 Lower Bound Estimator for Purely Algebraic Coupling

In practice, one is interested to measure the contraction factor during co-simulation procedure. Knowledge about this may be useful for an effective time window size control. For non-linear models the computation of $\mathbf{G}_{\mathbf{z}^{(k)}}^{-1}$, $\mathbf{G}_{\mathbf{z}^{(k-1)}}$ in (8.3) is costly. Within this chapter, we propose a lower bound estimation for α_n. The accuracy is analyzed in Sect. 8.4.

In multiphysics, the data exchange is frequently managed by algebraic constraints. Therefore, we investigate coupled DAE systems with the splitting functions

$$\mathbf{F}(\cdot,\cdot,\cdot,\cdot) = \begin{bmatrix} \mathbf{f}_1(\mathbf{y}_1,\mathbf{z}_1,0,0) \\ \mathbf{f}_2(0,0,\mathbf{y}_2,\mathbf{z}_2) \end{bmatrix}, \ \mathbf{G}(\cdot,\cdot,\cdot,\cdot) = \begin{bmatrix} \mathbf{g}_1(\mathbf{y}_1,\mathbf{z}_1,0,\mathbf{z}_2) \\ \mathbf{g}_2(0,\mathbf{z}_1,\mathbf{y}_2,\mathbf{z}_2) \end{bmatrix}. \quad (8.4)$$

Due to the fact that the information transport between the subsystems is organized by algebraic constraints, i.e., $\mathbf{G}_{\mathbf{z}^{(k-1)}} \neq 0$, a contraction factor, see (8.3), occurs. Now, let $\mathbf{X}_n(t)$, $\tilde{\mathbf{X}}_n(t)$ be two waveforms on the n-th time window H_n. The difference after k iterations is measured by $\delta_n^{(k)} := \|\mathbf{X}_n^{(k)}(t) - \tilde{\mathbf{X}}_n^{(k)}(t)\|_{2,\infty}$. Due to purely algebraic-to-algebraic coupling, i.e., the ODE parts in (8.4) are decoupled, the error propagation for the differential and algebraic part using a constant $C > 0$ reads:

$$\begin{pmatrix} \delta_{\mathbf{y},n}^{(k)} \\ \delta_{\mathbf{z},n}^{(k)} \end{pmatrix} \leq \mathbf{K}^k \begin{pmatrix} \delta_{\mathbf{y},n}^{(0)} \\ \delta_{\mathbf{z},n}^{(0)} \end{pmatrix} \quad \text{with} \quad \mathbf{K} = \begin{pmatrix} 0 & CH_n \\ 0 & CH_n + \alpha_n \end{pmatrix} \tag{8.5}$$

Thus we found that the contraction factor defined in (8.3) is bounded from below by

$$\alpha_n \geq \sqrt[k]{\delta_{\mathbf{z},n}^{(k)}/\delta_{\mathbf{z},n}^{(0)}} - CH_n. \tag{8.6}$$

Hence (8.6) enables us to estimate the contraction factor for H_n small enough. In practice, the error $\delta_n^{(k)}$ can be estimated via Richardson Extrapolation. To this end, the macro step is computed using the step size H_n and using two steps of size $H_n/2$. The comparison of both solutions $\mathbf{X}_n^{(k)}(t_n)$ and $\mathbf{X}_{n/2}^{(k)}$ gives the desired estimate:

$$\delta_{\cdot,n}^{(k)} = \frac{2^{p+1}}{2^{p+1} - 1} \|\mathbf{X}_n^{(k)}(t) - \mathbf{X}_{n/2}^{(k)}(t)\|_{2,\infty} + \mathcal{O}(H_n^{p+2}) \tag{8.7}$$

Here p denotes the degree of the polynomials, which are used to compute the initial guess $X_n^{(0)}$. Constant extrapolation, i.e., $p = 0$, is a common choice in co-simulation.

8.3 Density Estimation Techniques

When co-simulation is applied to systems with uncertain behavior, e.g., uncertain components in electrical circuits or production-related tolerances in electrical machines, it may influence the convergence, that is the contraction factor defined in (8.3) becomes stochastic with its own PDF denoted by p_{α_n}. Thus, the splitting (8.2) depends on Q uncertain parameters $\xi = (\xi_1, \ldots, \xi_Q)$ and all of them may affect the contraction behavior. Within this paper we only investigate (8.4) depending on one single uncertain parameter, i.e. $Q = 1$. Therefore, the objective is to estimate the PDF of the contraction factor as precise as possible. For that purpose several well known approaches can be found for measuring the PDF of random variables, see [4]. One prominent technique is the so-called Kernel Density Estimation (KDE) based on a brute-force sampling of the parameter space, see [5]. A different approach tries to determine the PDF analytically by covering the stochastic process using PC-expansion, see [6].

Both methods are fundamentally differently and provide specific benefits. To our knowledge, both have not been applied to co-simulation with respect to PDF measuring problems. Thus, a comparison of both methods in co-simulation is of interest.

Kernel Density Estimation Method In contrast to determine stochastic distributions by histograms, KDE allows to get PDFs without discontinuities and with fewer samples. Let $X = (x_1, \ldots, x_n) \in \mathbb{R}^n$ be an independent sample drawn from a distribution with unknown density p_X. Via KDE, the PDF can be estimated by

$$\hat{p}_X(x) = \frac{1}{nh} \sum_{i=1}^{n} K\left(\frac{x - x_i}{h}\right), \quad K(x) = \frac{1}{\sqrt{2\pi}} e^{-x^2/2} \tag{8.8}$$

where $K(x)$ is the Gaussian-kernel and $h > 0$ is a smoothing parameter, see [5]. To avoid oversmoothing, KDE requires a judicious choice of the bandwidth h. The parameter h can be chosen in an optimal way, see [4].

Spectral Method Let the approximation of the stochastic process of α_n be given by the PC-expansion

$$\alpha_{\mathrm{gPC}}(\xi) = \sum_{|\beta| \leq p} \alpha_\beta \Phi_\beta(\xi), \tag{8.9}$$

with multiindex $\boldsymbol{\beta} = (\beta_1, \ldots, \beta_Q) \in \mathbb{N}_0^Q$ thus that $|\boldsymbol{\beta}| := \beta_1 + \cdots + \beta_Q$ is the polynomial degree, the maximum polynomial degree p, coefficient functions α_β, multivariate polynomials $\Phi_\beta(\cdot)$ with $\langle \Phi_n(\cdot), \Phi_m(\cdot) \rangle = \delta_{nm}$, $n, m \in \mathbb{N}_0$ and $\boldsymbol{\xi}$ is random variable with probability density function $p_\xi : \mathbb{R}^Q \to \mathbb{R}_+$, see [7]. For simplicity, we abbreviate (8.9) as

$$X(\xi) = \alpha_{\mathrm{gPC}}(\xi). \tag{8.10}$$

To determine the coefficients α_β we employ stochastic collocation, see [7], which requires numerical quadrature to multidimensional integrals. Now we want to deduce a PDF for X, say $\hat{p}_X(\cdot)$. Suppose we want to evaluate this PDF at some x: Then

$$\hat{p}_X(x) = \sum_{\hat{\xi} \in R_x} \frac{p_\xi(\hat{\xi})}{\left| DX(\hat{\xi}) \right|} \quad \text{with} \quad R_x = \{\hat{\xi}_1, \ldots, \hat{\xi}_N\}, \tag{8.11}$$

where $\hat{\xi}_1, \ldots, \hat{\xi}_N$ are the N roots of the polynomial $X(\xi) - x = 0$, see [6]. In other words, many possible $\hat{\xi}$ may gives use this particular x and all of them contribute to the probability density at x. For one dimension, i.e., $X \colon \mathbb{R} \to \mathbb{R}$, the denominator becomes the absolute value of the first derivative: $|DX(\hat{\xi})| = |dX/d\xi$ evaluated at $\hat{\xi}|$.

Remark 1 If one generalizes (8.11) to multiple dimensions, i.e., more than one random variable ($Q > 1$), the quantity $|DX(\hat{\xi})|$ becomes the absolute value of the Jacobian determinant. Here, some further steps are needed for $\dim(X) \neq \dim(\xi)$, see [6]. Furthermore, the root-finding becomes much more difficult.

8.4   Numerical Test Example

Our test example serves a field/circuit coupled problem, [8], see Fig. 8.1. The data exchange between the subsystems is organized via source coupling, i.e., controlled current and voltage sources, see [2]. The coupling structure (8.4) is realized for this test example by using the novel technique of R-splitting, where the controlled sources are directly connected to resistances. In fact, the resistance R is split into two resistances in series: $R_1 := \omega R$, $R_2 := (1 - \omega)R$, $\omega \in [0, 1]$, see Fig. 8.1.

Assumption 1 (Avoiding Index-2) *It is assumed that the coupling sources are placed such that both (all) subsystems remain index-1 problem for all $\omega \in [0, 1]$, see [8]. This assumption is essential for the limits, i.e., $\omega = 0$ and $\omega = 1$.*

Employing the technique for the exact recursion deduction, see [2], the contraction factor and the constant read

$$\alpha_n = \frac{G_2}{G_1}, \quad C = G_2 \left(\frac{G_2}{G_1 [1 - L_F(1 + L_\Phi)H_n]} + \frac{\frac{G_2}{1 - L_F(1 + L_\Phi)H} H - 1}{1 - (L_F(1 + L_\Phi) + G_1)H} \right), \quad (8.12)$$

with Lipschitz constants L_Φ, L_F for the algebraic and differential splitting functions $\mathbf{G}$, $\mathbf{F}$ and conductances $G_1 := 1/R_1$, $G_2 := 1/R_2$. Thus, the strength of coupling is determined by the ratio of the coupling resistances. Furthermore, the constant C can be controlled by resistance splitting. That is, for increasing resistance R_2, the contraction factor α_n and the constant C decreases.

Remark 2 If $\omega = 0$ or $\omega = 1$, the split subsystems are coupled via the ordinary differential equations. Hence the structure (8.4) of the splitting functions is not given.

Speed of Convergence Now we choose the ratio parameter for R-splitting as $\omega = 0.05$ and $\alpha_n = 1/19$ occurs. Thus, a major part of resistance R is displaced to the field part. Our co-simulation is tested on $\Gamma = [0, 0.01]$ s using a constant $H_n = 10^{-3}$ s. The strong coupled system (8.1) is also computed on Γ and serves as our

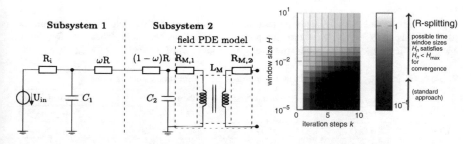

Fig. 8.1 (left) Field/circuit using R-splitting. Settings: $R_i = 10\Omega$, $R = 20\Omega$, no-load test, $C_1 = C_2 = 1$ nF, $U_{in}(t) = 170$ V $\cdot \cos(2\pi 60 \cdot t)$. (right) Window sizes H_n versus number of iterations k for cutting at the EM-Device boundaries (standard approach) and for R-splitting

reference solution. Both problems are solved using the MATLAB routine ode23s (Rosenbrock method) with accuracy AbsTOL = RelTOL = 10^{-5}.

Figure 8.2 compares the computational effort for both approaches, i.e., standard approach (when the computation starts with the field first) and R-splitting. Therefore, using R-splitting with small conductance G_2, the coupling becomes stronger and the number of iterations decreases for given accuracy, see Fig. 8.2 (left). This is why R-splitting only needs two iteration to be close to the reference solution. Hence, R-splitting converges faster and accepts larger time window sizes, see Fig. 8.1 (right).

In terms of solved linear systems the computational effort can be reduced about 90% to achieve the same accuracy in the solution, see Fig. 8.2 (right). A further reduction of the ratio parameter ω, i.e. $\omega = 0.005$, only leads to a small improvement.

Estimator Accuracy For our test example (Fig. 8.1), the contraction factor is given analytically, see (8.12). Thus the accuracy of (8.6) can be easily assessed. The simulation settings are the same as above.

Table 8.1 shows that the lower bound is fulfilled for each α_n, i.e., for each ratio between R_1 and R_2. However, the estimated values becomes less accurate for increasing value of α_n. This can be explained as follows: The contraction factor α_n and the constant C increases for increasing conductance G_2, i.e., the coupling becomes weaker. Thus the impact of the diagonal term CH_n in (8.5) is getting

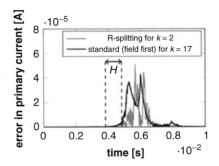
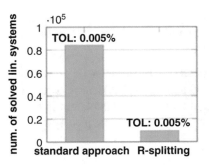

Fig. 8.2 (left) Error in the primary current of the transformer for co-simulation in the interval $[0, 0.01]$ s using $H = 10^{-3}$ s. (right) Computational effort for both approaches, i.e., R-splitting and cutting at the EM-Device boundaries (field first), in terms of solved linear systems

Table 8.1 Accuracy of the lower bound estimation for increasing values of α_n

Exact α_n	Estimated α_n	Abs. error	Window size H_n	Co-simulation
0.035	0.0502	0.017	$H = 10^{-3}$ s	Convergent
0.33	0.280	0.05	$H = 10^{-3}$ s	Convergent
0.96	0.472	0.488	$H = 10^{-3}$ s	Convergent
19	5.832	13.168	$H = 10^{-3}$ s	Divergent

more important. Consequently, using smaller step sizes H_n gives the opportunity to increase the accuracy of (8.6) even for larger contraction factors α_n.

KDE Versus Spectral Method In this section, we use the KDE and spectral method in order to estimate the PDF of the contraction factor. For one dimension, i.e., one random parameter in (8.4), the exact PDF p_{α_n} of the contraction factor $\alpha_n = \phi(\xi)$, where ξ is a random variable with PDF $p_\xi : [a, b] \rightarrow \mathbb{R}_+$, is given by the transformation formula for densities:

$$p_{\alpha_n}(\alpha_n) = p_\xi\left(\phi^{-1}(\alpha_n)\right)\left|\frac{d}{d\alpha_n}\phi^{-1}(\alpha_n)\right|, \quad \text{with } \phi: \mathbb{R} \rightarrow \mathbb{R}, \quad (8.13)$$

for details see [9]. A generalization to the multi-dimensional case is also given in [9].

For our test example, we focus on the one-dimensional case. We consider only R_1 or R_2 to be uniformly distributed $R_i \sim \mathscr{U}(10\,\Omega - \delta R_i, \ 10\,\Omega + \delta R_i)$, for $i = 1, 2$, with $\delta R_1 = 1\,\Omega$, $\delta R_2 = 7\,\Omega$. Generally, the mapping function ϕ is not explicitly known in co-simulation. However, using R-splitting, the function ϕ in (8.13) is given by (8.12). Now, the task is to estimate the PDF of α_n as good as possible from its underlying sample with as little effort as possible. Therefore, (8.13) enables to calculate the exact PDF of α_n and allows for a qualitative assessment of method (8.8) and (8.11).

Our testing works as follows: For each sample-point Ω, the reference model is solved in time domain up to $t_0 = 10^{-4}$s to obtain initial values which are close to the solution. The k iterations of the co-simulation are computed for each sample ω on $[t_0, t_0 + H_n]$. Next, we restart the computation using two steps with the half macro-step size, i.e., $[t_0, t_0 + H_{n/2}] \rightarrow [H_{n/2}, H_{n/2} + H_{n/2}]$. Therefore (8.7) can be used for error estimation in (8.6). Then, using the sample of α_n, we try to estimate the PDF by the KDE technique (8.8), and the spectral approach (8.11), where stochastic momenta (depending on step k) are computed. Furthermore, constant extrapolation of the initial value is used for the initial guess $\tilde{\mathbf{X}}_n^{(0)}(t)$ on time window H_n, $H_{n/2}$. Consequently (8.7) is of order $\mathcal{O}(H_n^2)$. For the spectral approach, the maximum degree of polynomial that we used for the approximation in (8.9) is two.

Figure 8.3 shows the estimated PDF using the spectral method and the KDE approach for different number of samples. The exact PDF is calculated by using (8.13). For R_1 uncertain, the spectral method becomes more accurate. In fact, due to the linear mapping, the spectral method enables to approximate the stochastic process exactly when at least linear polynomials in (8.9) are used. Then, the KDE approach needs a large number of samples to recover the uniform distribution precisely. However, for R_2 uncertain, the mapping function ϕ becomes non-linear, since R_2 in (8.12) is located in the denominator. Thus the spectral method is too coarse as long as we use quadratic polynomials for its approximation in (8.9). Using polynomials of higher order reduces the error. Figure 8.4 compares the accuracy of uniformly shaped PDF measured in the expectation value for increasing window sizes. Naturally, the error increases with increasing window size H_n. Hence, for H_n large the reduction per iteration is dominated by the second diagonal

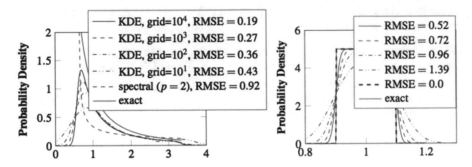

Fig. 8.3 PDF for uncertain resistance R_1, i.e., linear mapping function ϕ (right), and uncertain resistance R_2, i.e., non-linear mapping function ϕ (left), using spectral and KDE approach for increasing number of samples. The error is measured using the **R**oot **M**ean **S**quared **E**rror (RMSE)

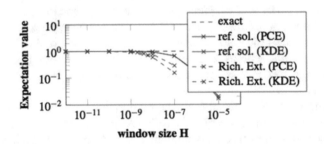

Fig. 8.4 Expectation value obtained by using KDE and spectral method with and without Richardson Extrapolation for different window sizes H

term CH_n. That is, (8.6) becomes less accurate. Furthermore, the inaccuracy of Richardson Extrapolation grows squarely with its applied window size. Using (8.7) as approximation for the exact solution, i.e., reference solution obtained by solving the strong coupled problem (8.1), yields approximately the same error as long as H_n is not too large. Therefore, using Richardson Extrapolation as error estimation in (8.6) seems to work precisely as long as we use window sizes $H_n \leq 10^{-9}$ s. There is the possibility to use larger window sizes if one is not interested in such high accuracy of about 10^{-4}.

8.5 Conclusions

We proposed a new approach for coupling (R-splitting), where the strength of coupling between the subsystems can be affected. With respect to the standard approach, i.e., cutting at the EM-Device boundary, we showed that R-splitting yields better properties regarding the speed of contraction.

Next, we showed the possibility to estimate the contraction factor online, i.e., during co-simulation, by using a lower bound. We used this to calculate a sample for measuring the underlying probability density function. This may also be of interest for an effective time window size control algorithm. However, further investigations are needed to find an upper bound, which enables to detect guaranteed convergence.

Furthermore, we demonstrated that the spectral- and the KDE approach are suitable techniques for estimating the distribution of the contraction factor in co-simulation. Knowledge about the distribution may help to guarantee convergence and contraction when co-simulation and uncertainties are combined. Particularly for many uncertain parameters, both schemes becomes very costly. In the future, we aim to reduce the computational effort. Therefore, calculation of sensitivity coefficients may help for effective parameter space sampling. This seems to be useful especially for the spectral approach, where the coefficient functions are already computed.

Acknowledgements This work is supported by the German Federal Ministry of Education and Research (BMBF) in the research projects SIMUROM (05M13PXB) and KoSMos (05M13PXA).

References

1. Arnold, M., Günther, M.: Preconditioned dynamic iteration for coupled differential-algebraic systems. BIT **41**, 1–25 (2001)
2. Bartel, A., Brunk, M., Günther, M., Schöps, S.: Dynamic iteration for coupled problems of electric circuits and distributed devices. SIAM J. Sci. Comput. **35**(2), B315–B335 (2013)
3. Burrage, K.: Parallel Methods for Systems of Ordinary Differential Equations. Clarendon Press, Oxford (1995)
4. Scott, D.W.: Multivariate Density Estimation: Theory, Practice, and Visualization, 2nd edn. Wiley, Hoboken (2015)
5. Kornatka, M.: The weighted kernel density estimation methods for analysing reliability of electricity supply. In: 17th International Scientific Conference on Electric Power Engineering (EPE), Prague, pp. 1–4 (2016)
6. Li, J., Marzouk, Y.: Multivariate semi-parametric density estimation using polynomial chaos expansion (2013, in preparation)
7. Xiu, D.: Numerical Methods for Stochastic Computations: A Spectral Method Approach. Princeton University Press, Princeton (2010)
8. Schöps, S., De Gersem, H., Bartel, A.: A cosimulation framework for multirate time integration of field/circuit coupled problems. IEEE Trans. Magn. **46**(8), 3233–3236 (2010)
9. Graybill, F.A., Mood, A.M., Boes, D.C.: Introduction to the Theory of Statistics, International 3rd revised edn. Mcgraw-Hill Education, New York (1974)

Chapter 9
Multirate DAE/ODE-Simulation and Model Order Reduction for Coupled Field-Circuit Systems

Christoph Hachtel, Johanna Kerler-Back, Andreas Bartel, Michael Günther, and Tatjana Stykel

Abstract Considering distributed and lumped electromagnetic effects in device simulation yields coupled field-circuit systems, which are high dimensional systems of partial-differential-algebraic equations. Moreover, such systems exhibit largely varying time scales and are difficult in the numerical handling. To exploit the different dynamical behaviour of circuit and field equations, we propose multirate time integration schemes which are extended to differential-algebraic equations. These schemes are also combined with model reduction of a slow changing subsystem of magneto-quasistatic equations which significantly decreases the computational effort.

9.1 Introduction

For the development of modern electrical devices, the influence of electromagnetic effects has to be considered in the simulation process very often. In general, this leads to a coupled problem where the subsystems provide a quite different behaviour. In magneto-quasistatic (MQS) problems, the electromagnetic field is described by Maxwell's equation in the magnetic potential formulation

$$\sigma \frac{\partial A}{\partial t} + \nabla \times (\nu \nabla \times A) = J \quad \text{in } \Omega \times (0, T) \tag{9.1}$$

with appropriate boundary and initial conditions, where Ω is a bounded two- or three-dimensional domain composed of conducting and nonconducting

C. Hachtel (✉) · A. Bartel · M. Günther
Bergische Universität Wuppertal, Wuppertal, Germany
e-mail: hachtel@math.uni-wuppertal.de; bartel@math.uni-wuppertal.de;
guenther@math.uni-wuppertal.de

J. Kerler-Back · T. Stykel
Universität Augsburg, Institut für Mathematik, Augsburg, Germany
e-mail: kerler@math.uni-augsburg.de; stykel@math.uni-augsburg.de

© Springer International Publishing AG, part of Springer Nature 2018
U. Langer et al. (eds.), *Scientific Computing in Electrical Engineering*,
Mathematics in Industry 28, https://doi.org/10.1007/978-3-319-75538-0_9

subdomains, A is the magnetic vector potential, v is the magnetic reluctivity which may depend nonlinearly on A on the conducting subdomain, σ is the electric conductivity vanishing on the non-conducting subdomain, and J is the current density applied by external sources. Using modified nodal analysis, electric networks with distributed MQS devices can be modelled by a system of differential-algebraic equations (DAEs)

$$E(y)\tfrac{d}{dt}y = f(t, y, i_M),\tag{9.2}$$

where y contains all node potentials and currents through flux and voltage controlled elements [1]. These equations are coupled to the MQS equation (9.1) via the vector of lumped currents i_M through the distributed MQS devices. Taking $J = \chi i_M$ with a divergence-free winding function χ, the coupling equation connecting Maxwell's equation (9.1) to the network equation (9.2) is given by

$$\int_\Omega \chi^T \tfrac{\partial}{\partial t} A \, d\xi + R \, i_M = u,\tag{9.3}$$

where R is the resistance matrix and u is the vector of applied voltages.

Often, the network equations provide a faster dynamic behaviour than Maxwell's equation for the MQS devices. Such coupled systems can be solved efficiently by multirate time integration schemes, where the slow changing components are integrated with large macro-step sizes, while the fast changing components are integrated with small micro-step sizes. For systems of ordinary differential equations (ODEs), there are different approaches how the coupling between the subsystems can be realised, e.g. [2–4].

The novelty of the paper is twofold. First we extend the multirate concept of Savcenco et al. [4] to systems consisting of a fast changing subsystem of ODEs and a slow changing subsystem of DAEs. This method can be used for an ODE system describing an electrical circuit (9.2) after an index reduction and a DAE system obtained by a spatial discretisation of Maxwell's equation (9.1). Such a coupled system has high dimension and is time consuming in simulation. To decrease the computational effort, model order reduction is combined with the multirate time integration scheme. This is the second novelty of the paper. For model reduction of the semidiscretised MQS equations, we use a method which was developed in [5], this method starts with a full-order DAE system and ends up with a reduced-order ODE system.

The outline of the paper is as follows: First, we present a balanced truncation based model order reduction technique for the DAE formulation of a MQS equation which provides a reduced-order model in ODE form. Next, we introduce a multirate time integration scheme for a coupled system that consists of a fast changing ODE subsystem and a slow changing DAE subsystem. Finally, we apply the multirate time integration scheme combined with model reduction to the MQS equation for a single-phase 2D transformer embedded in an electrical circuit and present some results of numerical experiments.

9.2   Model Order Reduction for Magneto-Quasistatic Equations

In this section, we briefly discuss model order reduction of the MQS equations. For more details, we refer to [5]. Applying the finite element discretisation method to (9.1) and (9.3), we obtain a nonlinear system of DAEs

$$
\mathfrak{M}\frac{d}{dt}\begin{bmatrix} a \\ i_M \end{bmatrix} = \mathfrak{F}(a)\begin{bmatrix} a \\ i_M \end{bmatrix} + \mathfrak{B}\,u, \qquad w = \mathfrak{B}^T\begin{bmatrix} a \\ i_M \end{bmatrix} \tag{9.4}
$$

with a singular mass matrix $\mathfrak{M}$, a semidiscretized vector of magnetic potentials a, an input u and an output $w = i_M$. The properties of the involved system matrices guarantee that (9.4) is of index one and it can be transformed into a system of ODEs

$$
M\frac{d}{dt}z = F(z)z + Bu, \qquad w = -B^T M^{-1} F(z)z \tag{9.5}
$$

with a nonsingular matrix M and a corresponding vector of unknowns $z = z(t)$. Note that system (9.5) has the same input u and the same output w as the DAE system (9.4) meaning that the input-output relation of (9.4) is preserved in (9.5).

If the magnetic reluctivity is constant on the conducting domain, then $F(z)$ in (9.5) is independent of z resulting in a linear time-invariant system

$$
M\dot{z} = Fz + Bu, \qquad w = -B^T M^{-1} Fz \tag{9.6}
$$

with the symmetric, positive definite matrices M and $-F$ [5]. These conditions guarantee that (9.6) is asymptotically stable and passive. For model reduction of (9.6), we use a balanced truncation approach based on the controllability Gramian P which is defined as a unique symmetric and positive semidefinite solution to the generalized Lyapunov equation

$$
FPM + MPF = -BB^T. \tag{9.7}
$$

Due to the symmetry conditions, the observability Gramian Q satisfies $MQM = FPF$. Let $P = SS^T$ be a Cholesky factorization of P. We compute the eigenvalue decomposition

$$
-S^T FS = [\,U_1,\ U_0\,]\,\mathrm{diag}(\Lambda_1,\ \Lambda_0)\,[\,U_1,\ U_0\,]^T,
$$

where Λ_1 and Λ_0 are diagonal matrices and Λ_1 contains all kept Hankel singular values and Λ_0 all truncated ones. Now, we can determine the reduced-order model by projection

$$
\tilde{M}\dot{\tilde{z}} = \tilde{F}\tilde{z} + \tilde{B}u, \qquad \tilde{w} = \tilde{C}\tilde{z}, \tag{9.8}
$$

where $\tilde{M} = W^T M V$, $\tilde{F} = W^T F V$, $\tilde{B} = W^T B$ and $\tilde{C} = -B^T M^{-1} F V$ with the projection matrices $V = S U_1 \Lambda_1^{-1/2}$ and $W = -M^{-1} F V$. One can show that the reduced matrices $\tilde{M}$ and $-\tilde{F}$ are symmetric, positive definite and $\tilde{C} = \tilde{B}^T$ guarantees that system (9.8) is passive. Moreover, we have the L_2-norm error bound for the output

$$\|w - \tilde{w}\|_2 \leq 2 \operatorname{trace}(\Lambda_0) \|u\|_2.$$

For solving the generalized Lyapunov equation (9.7), we can use the low-rank alternating direction implicit method or (rational) Krylov subspace method [6, 7]. In both methods, we need to solve linear systems of the form $(\tau M + F)v = b$ for a vector v with possibly dense M and F. Exploiting the block structure of these matrices, we can overcome this computational difficulty by solving linear systems $(\tau \mathfrak{M} + \mathfrak{F})\hat{v} = \hat{b}$ with the sparse matrices $\mathfrak{M}$ and $\mathfrak{F}$ as in (9.4) instead [5].

For model reduction of the nonlinear system (9.5), we can use the proper orthogonal decomposition technique combined with the discrete empirical interpolation method (DEIM) for efficient evaluation of the nonlinearity $g(z) = F(z)z$ and matrix DEIM for fast computation of the Jacobi matrix $J_g(z)$, see [5] for details.

9.3   Multirate Time Integration for ODE/DAE-Systems

Now, we present an efficient time integration scheme to simulate electromagnetic effects in electrical devices. For the MQS equations, we consider the semidiscretised DAE formulation (9.4) and set $x = [a^\top, i_M^\top]^\top$. We claim that the surrounding electrical circuit can be described by a system of ODEs and its solution is denoted by y. Then the coupled system of equations reads:

$$\dot{y} = f(t, y, x) \tag{9.9}$$

$$\mathfrak{M}\dot{x} = \mathfrak{F}(x)x + \mathfrak{B}u. \tag{9.10}$$

The coupling from the ODE to the DAE is realised by the input function $u = u(y)$ in (9.3). The network ODE provides a faster dynamic behaviour than the DAE model of Maxwell's equations. Since the DAE (9.10) is a result from a finite elements semi-discretisation its dimension is much larger than the dimension of the circuit's ODE system. However, the coupled system can be written in the form of one DAE

$$G(t, \dot{y}, \dot{x}, y, x) = 0. \tag{9.11}$$

For given input u, it was shown in [5] that the DAE (9.10) is of tractability index 1. Thus the DAE (9.11) is also of index 1 and therefore it can be integrated by an implicit Runge-Kutta method [8]. We apply the LobattoIIIC method to this DAE with given consistent initial values $y(t_0) = y_0$, $x(t_0) = x_0$. For the first time step $t_0 \rightarrow t_0 + H$ this reads for the increments k_1^y, k_2^y, k_1^x, k_2^x as

$$G(t_0, \, k_1^y, \, k_1^x, \, y_0 + \tfrac{H}{2}(k_1^y - k_2^y), \, x_0 + \tfrac{H}{2}(k_1^x - k_2^x)) = 0,$$
$$G(t_0 + H, \, k_2^y, \, k_2^x, \, y_0 + \tfrac{H}{2}(k_1^y + k_2^y), \, x_0 + \tfrac{H}{2}(k_1^x + k_2^x)) = 0. \tag{9.12}$$

System (9.12) has to be solved with respect to k_1^y, k_2^y, k_1^x and k_2^x. Then, the approximations for y and x at $t_0 + H$ are given by

$$y_H = y_0 + \tfrac{H}{2}(k_1^y + k_2^y), \tag{9.13}$$

$$x_H = x_0 + \tfrac{H}{2}(k_1^x + k_2^x) \tag{9.14}$$

Here, the fast changing ODE subsystem dictates the step size H for the whole coupled DAE (9.11). This leads to a large computational effort since the whole high dimensional system has to be integrated with relatively small step sizes to resolve the network dynamics appropriately and it makes the time domain simulation of the coupled system inefficient.

A multirate time integration scheme decreases the computational effort and preserves the accuracy of the numerical approximation. The slow changing subsystem (9.10) is integrated with a large macro-step size H while the fast changing subsystem (9.9) is integrated with a small micro-step size $h \ll H$. The crucial part is how the unknown function values of x at the intermediate time steps are achieved. For coupled systems of ODEs there are several approaches based upon inter- and extrapolation of the unknown values [2] or modified Runge-Kutta methods with inherent time steps for the coupled system [3].

Here we follow the idea of [4] and extend this technique to coupled ODE/DAE systems. First, the system (9.12) is solved for the overall coupled system (9.11) with macro-step size H which is chosen according to the system properties of the slow changing DAE subsystem (9.10). The approximation at $t_0 + H$ is only accepted for the slow changing subsystem according to (9.14) since an approximation with stepsize H for the fast changing ODE subsystem is inaccurate.

Now, the fast changing ODE subsystem (9.9) is integrated with a smaller micro-step size h over the time interval $[t_0, \, t_0+H]$. The system for the increments $k_1^{y,h} \, k_2^{y,h}$ of the first micro step $t_0 \rightarrow t_0 + h$ reads

$$k_1^{y,h} = f(t_0, \, y_0 + \tfrac{h}{2}(k_1^{y,h} - k_2^{y,h}), \, \bar{x}_1),$$
$$k_2^{y,h} = f(t_0 + h, \, y_0 + \tfrac{h}{2}(k_1^{y,h} + k_2^{y,h}), \, \bar{x}_2),$$

where $\bar{x}_1$ and $\bar{x}_2$ denote linearly interpolated values of x_0 and x_H at time t_0 and $t_0 + h$, respectively. The approximation of y at $t_0 + h$ is given by

$$y_h = y_0 + \tfrac{h}{2}(k_1^{y,h} + k_2^{y,h}). \tag{9.15}$$

The micro-step size h has to be chosen according to the dynamical properties of the solution y (fast circuit subsystem). It is also possible to include a step size control by embedding a lower order method. For LobattoIIIC the lower order approximation $\bar{y}_h$ can be computed by $\bar{y}_h = y_0 + hk_1^{y,h}$. After a certain number of micro-steps, an approximation $\hat{y}_H$ of y at $t_0 + H$ is achieved and the next macro-step $t_0 + H \rightarrow t_0 + 2H$ can be computed with corresponding initial values $\hat{y}_H$ and x_H as described above.

In case of a coupled field-circuit system, the fast changing circuit subsystem (9.9) depends on the current i_m, which is an algebraic variable of the slow changing MQS subsystem (9.10). To compute the micro-steps of the fast changing subsystem (circuit), an interpolation of i_m is needed. However, on the macro-step scale, the LobattoIIIC is stiffly accurate, thus the algebraic constraints will be satisfied.

9.4   Simulation of a Coupled Electric Field-Circuit System

We simulate the electromagnetic effects of a single-phase 2D transformer in a coupled field-circuit system. Since the transformer does not react immediately on fast changes in the input voltage, this system suits for integration by a multirate scheme. The fast changing subsystem describes the circuit, while the slow subsystem is used to model the electromagnetic effects of the transformer. Figure 9.1 shows a circuit diagram of the coupled system, where the electromagnetic effects are represented by the lumped devices of a transformer in the box.

MQS-Device Modeling   We consider the linear MQS equations for a single-phase 2D transformer with an iron core and two coils in the form (9.4). The material parameters are $\sigma = 5 \cdot 10^5 \Omega^{-1}\,\mathrm{m}^{-1}$, $\nu_1 = 14{,}872\,\mathrm{Am/(Vs)} = 14{,}872\,\mathrm{m/H}$ on the conducting and $\nu_2 = 1\,\mathrm{Am/(Vs)} = 1\,\mathrm{m/H}$ on the non-conducting subdomain. The FEM discretisation is done by the free available software FEniCS.[1] To apply a time domain simulation, the system matrices of the semidiscretised MQS system (9.4) of dimension $n_L = 7823$ were exported to MATLAB. The input of the subsystem is given by the voltage u at the primary coil, and the output is the current i_M through

Fig. 9.1 Circuit diagram for no load test of the coupled systems with lumped elements for the electromagnetic effects (box)

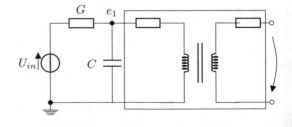

the primary coil. The reduced model was computed by the balanced truncation method as described in Sect. 9.2. The dimension of the reduced model is $r = 4$.

Circuit Modeling and Coupling The electric circuit and the transformer are coupled by the source coupling approach [9]. That is, add an additional controlled current source to the circuit subsystem and an additional voltage source to the transformer's subsystem. In this case, the circuit is described by the following ODE

$$C\frac{d}{dt}e_1(t) = G(e_1(t) - U_{in}(t)) - i_M(t)$$

for the node potential e_1, while i_M denotes the coupling current (as controlled current source) through the primary coil of the transformer. The circuit parameters are given by $C = 1\,\text{nF}$ and $G = 10^{-3}\,\text{S}$. The input voltage is given by two superposed sine functions $U_{in}(t) = 45.5 \cdot 10^3 \sin(900\pi t) + 10^3 \sin(45000\pi t)$, and the output is e_1.

Simulation Results We integrate the system by the multirate LobattoIIIC scheme over the time interval [0 s, 0.0055 s] as described in Sect. 9.3. Since we are interested in the influence of the multirate approach, we consider a reference solution that is computed by the LobattoIIIC method with constant global step size using 2500 time steps. We also integrated the coupled system with constant global step size using the double amount of time steps. The maximum relative 2-norm error in the outputs of the subsystems between both solutions was $3.9 \cdot 10^{-3}$. So we accepted the 2500 time step solution as reference solution with a moderate accuracy. The simulation was run on a Intel Core2 Duo P7450 with 2.13 GHz with 4 GB RAM. For the coupled DAE/ODE system of full-order, the computation time was 728.2 s. Figure 9.2 shows the outputs of the two subsystems: (a) the node potentials e_1, which belongs to the fast changing subsystem (basically we see the superposition of the sinusoidal oscillations) and (b) the current through the primary coil of the transformer, which belongs to the slow subsystem.

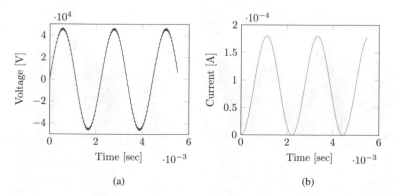

Fig. 9.2 Numerical solution of the subsystems. (**a**) Node potential of e_1. (**b**) Current through the primary coil

To investigate the influence of the multirate approach on the full order DAE system, the time interval is discretised into 250 macro-step and each macro-step is refined into 10 micro-steps. 250 macro-steps are sufficient to integrate the slow changing field subsystem and 2500 micro-steps are needed for the fast changing circuit subsystem to reach an adequate approximation. Here the computation ended after 77.4 s. We computed the error between the single-rate reference solution and the multirate approximation separately for both subsystems. For the fast changing subsystem, the error is computed by the absolute value of the difference between the node potential of the reference solution and the node potential achieved by the multirate approximation at each micro-step. For the slow changing subsystem, we computed the absolute value of the difference in the output of the subsystem i_M at the macro-steps. Figure 9.3 illustrates these errors. In the fast changing subsystem the error increases during one macro-step since there is an additional error that is caused by interpolating the values of the slow changing subsystem. At the macro-steps the subsystems are integrated together, so that the error at these time points is usually a bit smaller. In the slow subsystem, every second approximation gives better results while the intermediate approximation is worse. Until now, this phenomena is not yet understood completely. Since the size of the error is in total small, the improvement in computation time motivates and justifies the usage of multirate time integration schemes for these DAEs.

The reduced-order coupled system is integrated by the same multirate method with the same integration parameters as for the full-order system. The simulation needed 0.20 s to compute. Figure 9.4 shows the absolute error between both multirate approximations. The error here is very small and fits to the error bound results of [5].

Finally, we integrated the coupled system with the reduced MQS subsystem (9.8) without multirating, so we used the same integration parameters as for the DAE reference solution. The computation time was 0.13 s, so it was a bit faster than with multirating. This phenomena can be explained by the ratio between the number of

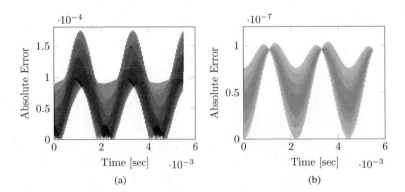

Fig. 9.3 Absolute error between multirate and singlerate approximations. (**a**) Circuit subsystem. (**b**) Electromagnetic subsystem

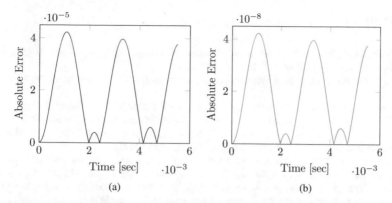

Fig. 9.4 Absolute errors in the subsystems resulting from model reduction of the MQS subsystem. (**a**) Circuit subsystem. (**b**) Electromagnetic subsystem

fast and slow changing variables. In our case, the full-order system has a ratio of 1 : 7821, while for the reduced-order system, it is 1 : 4.

This ratio is an indicator for the gain of efficiency between the singlerate and multirate approximation. If there is a large number of slow changing variables compared to a small number of fast changing variables, a multirate time integration scheme saves many function evaluation of the large dimensional slow subsystem. However, the implementation of a multirate scheme is more complex than for a classical singlerate scheme. So if the dimension of the slow changing subsystem is only a little bit larger than the dimension of the fast changing subsystem, a multirate scheme can be even less efficient than the corresponding singlerate scheme.

9.5   Conclusions

We combined two approaches for an efficient simulation of coupled circuit-field systems. By extending multirate time integration to DAE systems, these schemes can be applied to a larger class of problems to reduce the computation time significantly. Model order reduction for MQS equations decreases further the computational effort and the numerical handling is much easier since we only have to deal with a system of ODEs. Both approaches and their combination provide reliable approximations with small errors. We pointed out that the efficiency of multirate time integration schemes strongly depends on the ratio between the number of fast and slow changing variables. The combination of model order reduction and multirate time integration is advisable for systems where the dimension of the reduced order subsystem remains high compared to the dimension of the fast changing subsystem.

Acknowledgements This work was supported by the Research Network KoSMos: *Model Reduction Based Simulation of Coupled PDAE Systems* funded by the German Federal Ministry of Education and Research (BMBF), grants no. 05M13PXA and 05M13WAA. Responsibility for the contents of this publication rests with the authors.

References

1. Günther, M., Feldmann, U.: CAD based electric circuit modelling I: impact of network structure and parameters. Surv. Math. Ind. **8**, 97–129 (1999)
2. Gear, C.W., Wells, D.R.: Multirate linear multistep methods. BIT Numer. Math. **24**(4), 484–502 (1984)
3. Kværnø, A., Rentrop, P.: Low order multirate Runge-Kutta methods in electric circuit simulation (1999). Norges teknisk-naturvitenskapelige universitet Trondheim, Preprint No. 2/99
4. Savcenco, V., Hundsdorfer, W., Verwer, G.J.: A multirate time stepping strategy for stiff ordinary differential equations. BIT Numer. Math. **47**(1), 137–155 (2007)
5. Kerler-Back, J., Stykel, T.: Model order reduction for linear and nonlinear magneto-quasistatic equations. Int. J. Numer. Methods Eng. (2017, to appear). https://doi.org/10.1002/nme.5507
6. Benner, P., Saak, J.: Numerical solution of large and sparse continuous time algebraic matrix Riccati and Lyapunov equations: a state of the art survey. GAMM-Mitteilungen **36**(1), 32–52 (2013)
7. Simoncini, V.: Computational methods for linear matrix equations. SIAM Rev. **58**(3), 377–441 (2016)
8. Brenan, K., Campbell, S., Petzold, L.: Numerical Solution of Initial-Value Problems in Differential-Algebraic Equations. Classics in Applied Mathematics, vol. 14. SIAM (1995)
9. Bartel, A., Brunk, M., Günther, M., Schöps, S.: Dynamic iteration for coupled problems of electric circuits and distributed devices. SIAM J. Sci. Comput. **35**(2), B315–B335 (2013)

Chapter 10
Modelling and Simulation of Electrically Controlled Droplet Dynamics

Yun Ouédraogo, Erion Gjonaj, Thomas Weiland, Herbert De Gersem,
Christoph Steinhausen, Grazia Lamanna, Bernhard Weigand,
Andreas Preusche, and Andreas Dreizler

Abstract The electrohydrodynamics of millimetric droplets under the influence of slowly varying electric fields is considered. Strong electric fields applied on liquids induce forces driving fluid motion. This effect can be used, among others, in on-demand droplet generators. In this work, we discuss a convection-conduction model for the simulation of droplet motion in strong electric fields. The model focuses on robustness with respect to topology changes and on dynamic charging effects in liquids. We illustrate the model with the simulation of electrically driven droplet generation. The simulated dynamics for droplets with different conductivities are compared with experiments.

10.1   Introduction

The application of strong electric fields on liquids is used in many engineering applications to induce liquid atomization in a controlled manner. In electrosprays, the droplet size and the opening angle of the spray cone can be affected by charging the liquid prior to its atomization. Electric fields can also be used to

Y. Ouédraogo (✉) · E. Gjonaj · T. Weiland · H. De Gersem
Technische Universität Darmstadt, Institut für Theorie Elektromagnetischer Felder, Darmstadt, Germany
e-mail: ouedraogo@temf.tu-darmstadt.de; gjonaj@temf.tu-darmstadt.de;
thomas.weiland@temf.tu-darmstadt.de; degersem@temf.tu-darmstadt.de

C. Steinhausen · G. Lamanna · B. Weigand
Universität Stuttgart, Institut für Thermodynamik der Luft- und Raumfahrt, Stuttgart, Germany
e-mail: christoph.steinhausen@itlr.uni-stuttgart.de; grazia.lamanna@itlr.uni-stuttgart.de;
bernhard.weigand@itlr.uni-stuttgart.de

A. Preusche · A. Dreizler
Technische Universität Darmstadt, Fachgebiet für Reaktive Strömungen und Messtechnik, Darmstadt, Germany
e-mail: preusche@rsm.tu-darmstadt.de; dreizler@rsm.tu-darmstadt.de

© Springer International Publishing AG, part of Springer Nature 2018
U. Langer et al. (eds.), *Scientific Computing in Electrical Engineering*,
Mathematics in Industry 28, https://doi.org/10.1007/978-3-319-75538-0_10

achieve controlled motion of single droplets. In on-demand droplet generators, detachment induced by electric fields allows reproducible generation of liquids samples for experiments in various atmospheric conditions [1]. The modelling of electrically driven droplet motion involves a mechanical and an electrical problem. The externally applied electric fields induce forces on the droplets, driving fluid flow. The motion the phase boundaries affects the fields in turn, resulting in a strong coupling between the two problems. Furthermore, due to the presence of intrinsic ionic species and dissolved impurities, liquids exhibit some electrical conductivity associated with charge migration [2, 3]. Free charge accumulate at the droplet interface and can result in the generation of charged droplets from initially uncharged liquid. Such a leaky dielectric behaviour requires an electroquasistatic field representation, taking into account both conduction and displacement electric currents in the liquid.

In this work, we discuss a conduction-convection model for the simulation of droplet dynamics under the influence of electric fields, using the Finite Volume Method on a fixed mesh. The model accounts for electroquasistatic droplet charging, contact angle dynamics, and phase boundary topology changes associated with the droplet detachment process. The simulation of controlled droplet generation in an electrically driven droplet generator is presented and compared with experimental data.

10.2   Numerical Model

In the presence of electric fields, fluid motion induced by the electric forces distort the phase boundaries. The electric field distribution is affected in turn by changes in the free charge distribution and interface motion. The flow is therefore determined by solving in a coupled manner the multiphase hydrodynamic problem and the electroquasistatic problem.

10.2.1   Fluid Problem

In this work, the electrohydrodynamic problem is solved on a fixed grid using a Finite Volume discretization based on the OpenFOAM simulation framework [4]. The incompressible Navier-Stokes equations are adopted to describe fluid motion:

$$\frac{\partial \hat{\rho} \mathbf{u}}{\partial t} + \nabla \cdot (\hat{\rho} \mathbf{u} \mathbf{u}) - \hat{\mu} \nabla \cdot \left(\nabla \mathbf{u} + \nabla \mathbf{u}^{\mathrm{T}} \right) = -\nabla p + \hat{\rho} \mathbf{g} + \mathbf{f}_{\mathrm{s}} + \mathbf{f}_{\mathrm{e}} , \tag{10.1}$$

$$\nabla \cdot \mathbf{u} = 0 , \tag{10.2}$$

where $\mathbf{u}$, p, $\hat{\rho}$ and $\hat{\mu}$ are, respectively, the fluid velocity, dynamic pressure, density and the viscosity of the fluids. The external forces, $\mathbf{f}_e$, and $\mathbf{f}_s$, are respectively the electric forces applied to the fluids, and the surface tension force. The material properties $\hat{\rho}$ and $\hat{\mu}$ are locally averaged material properties, calculated using the Volume of Fluid approach representing the interface between fluids, further described in Sect. 10.2.2.

Surface wettability is accounted for by a contact angle model, providing a correction for the surface tension force $\mathbf{f}_s$ on the first boundary layer. In this work, the Kistler correlation is used, representing the dynamic contact angle θ as a function of the capillary number Ca at the contact line [5]:

$$\theta = f_H \left(Ca + f_H^{-1} \left(\theta_{adv/rec} \right) \right), \tag{10.3}$$

$$\text{with } f_H(x) = \arccos \left(1 - 2 \tanh \left[5.16 \left(\frac{x}{1 + 1.31x^{0.99}} \right)^{0.706} \right] \right) \tag{10.4}$$

In (10.3), θ_{adv} and θ_{rec} represent, respectively, the limiting advancing and receding angles of the liquid on the given surface. Pinning effects on the contact line are furthermore modelled by keeping the volume fraction constant at boundaries where the apparent angle value is between the critical contact angle values.

10.2.2   Interface Capturing

The Volume of Fluid (VoF) approach [6] is the method of choice for the simulation of multiphase fluid problems. It allows for a numerically efficient representation of the phase boundaries, in particular when topology changes such as droplet breakup and collision are involved. Phase boundaries are captured on the fixed grid using the volume fraction, α, occupied by one of the fluids in each cell of the computational mesh. Local material properties at the interface are defined from the phase fraction by weighted averaging between the properties of the involved fluids:

$$\hat{\mu} = \alpha \mu_{liq} + (1 - \alpha)\mu_{gas}, \quad \hat{\rho} = \alpha \rho_{liq} + (1 - \alpha)\rho_{gas}, \tag{10.5}$$

where μ and ρ denote the dynamic viscosity and the density of the fluids, respectively. The multiphase problems are then solved considering the two phases as a single phase with varying material properties. The surface force densities in (10.1) therefore become volume force density, applied across the diffuse interface. The evolution of the interface is characterized by the transport equation of the volume fraction:

$$\frac{\partial \alpha}{\partial t} + \nabla \cdot ([\mathbf{u} + \alpha(1 - \alpha)\mathbf{u}_c] \alpha) = 0, \tag{10.6}$$

where $\mathbf{u_c}$ is an artificial compression velocity acting as a counter diffusive term. Equation (10.6) ensures mass conservation. Moreover, topology changes in the phase boundaries are implicitly handled.

10.2.3 Electric Problem

The electric fields involved in the dynamics of droplet charging are essentially irrotational, and are determined by:

$$\nabla \cdot (\varepsilon \mathbf{E}) = \rho_e, \ \mathbf{E} = -\nabla \Phi. \tag{10.7}$$

The electroquasistatic behaviour is represented via the charge conservation equation:

$$\frac{\partial \rho_e}{\partial t} + \nabla \cdot \mathbf{J} = 0, \tag{10.8}$$

where $\mathbf{J} = \rho_e \mathbf{u} + \kappa \mathbf{E}$ is the current density including the ohmic conduction in the fluids and the convection of free charge in the liquids. The transient electroquasistatic field is finally given by:

$$\nabla \cdot \hat{\varepsilon} \nabla \Phi = -\rho_e, \tag{10.9}$$

$$\frac{\partial \rho_e}{\partial t} + \nabla \cdot (\rho_e \mathbf{u}) = \nabla \cdot \hat{\kappa} \nabla \Phi. \tag{10.10}$$

The VoF approach is used in this work also for representing electrical material properties on the grid. This approach provides better numerical accuracy than a simple staircase approximation of the phase boundary. The electric permittivity, $\hat{\varepsilon}$, and conductivity, $\hat{\kappa}$, used in (10.9) and (10.10) are therefore defined, similarly to (10.5), as:

$$\frac{1}{\hat{\varepsilon}} = \frac{\alpha}{\varepsilon_{liq}} + \frac{1-\alpha}{\varepsilon_{gas}}, \ \frac{1}{\hat{\kappa}} = \frac{\alpha}{\kappa_{liq}} + \frac{1-\alpha}{\kappa_{gas}}, \tag{10.11}$$

where a harmonic averaging of electrical material properties based on the phase fraction, α, is applied [7, 8].

Given a numerical solution for (10.9), (10.10), volumetric forces applied on the liquid can be calculated from a modified Maxwell stress tensor, holding for incompressible fluid flow:

$$\mathbf{f}_e = \nabla \cdot \left(\hat{\varepsilon} \mathbf{E} \otimes \mathbf{E} - \frac{1}{2} \varepsilon_0 E^2 \mathbf{I} \right). \tag{10.12}$$

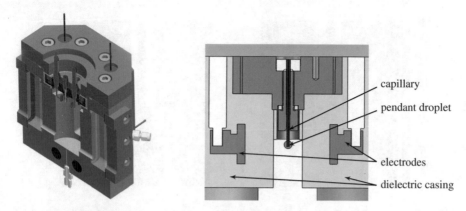

Fig. 10.1 Schematics of the on-demand droplet generator, cf. [1]. Left: overview of the pressure chamber. Right: capillary and electrodes triggering the detachment

10.3   Application and Results

The electrohydrodynamic model is applied in the simulation of the droplet generator shown in Fig. 10.1. A metallic capillary tube at ground voltage is placed between two electrodes in a climate chamber capable of sustaining a high-pressure, high-temperature gaseous environment. Liquid is introduced from the top of the capillary, forming a droplet that wets the sides of the capillary. As the droplet reaches a specified size, a short voltage pulse is applied at the electrodes. The electric forces induced on the droplet accelerate it for the duration of the pulse.

10.3.1   Electrically Driven Droplet Detachment

The mechanism involved in droplet detachment differ depending on the ratio between the free charge relaxation time in the considered liquids and the duration of the applied voltage pulse. The electric field and force distributions around liquid droplet are shown in Fig. 10.2. For liquids with a short relaxation time, in the absence of electric fields inside of the liquid, the electric field maximum is at the bottom of the droplet. As a result, the electric forces provides a downward impulse to the droplet. The electric pulse is interrupted before detachment, to prevent from detaching charged droplets. In low conductivity liquids, however, the electric field maximum is at the tip of the capillary, regardless of the position of the droplet. The resulting force distribution provides an upward impulse to the droplet. After the end of the electric pulse, the weight of the droplet, now uncompensated by the surface tension provides the downward acceleration leading to detachment.

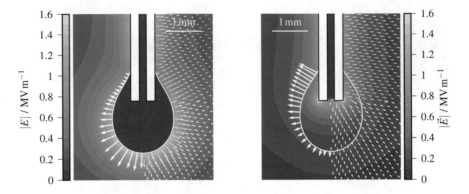

Fig. 10.2 Electric fields and forces distributions acting on acetone and n-pentane droplets. Left: acetone droplet; Right: n-pentane droplet. The left part of each graphs shows the electric force distribution, represented as a surface force density

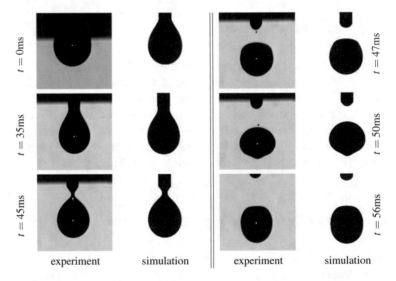

Fig. 10.3 Comparison between the simulated and experimentally recorded droplet dynamics in the generator during the detachment process, for acetone at 20 °C, 1 bar, using a 2 kV, 11.5 ms voltage pulse. Left: experiment; Right: simulation

In order to reduce the computational cost, the droplet generator is modelled assuming axial symmetry. The contact angle parameters on the capillary are empirically chosen such as to reproduce the droplet shape shortly before the voltage pulse is applied.

The simulated dynamics of detachment are compared in Figs. 10.3 and 10.4, for acetone and n-pentane droplets, respectively. A good agreement between simulation and experiment can be observed. Small deviations in the computed droplet shapes

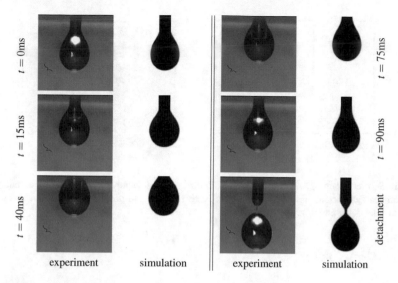

Fig. 10.4 Comparison between the simulated and experimentally recorded droplet dynamics in the generator during the detachment process, for n-pentane at 27 °C, 1 bar, using a 4 kV, 26.5 ms voltage pulse. Left: experiment; Right: simulation

are most probably due to uncertainties related to the determination of capillary surface wettability parameters and uncertainties on material parameters.

10.3.2 Charged Droplet Detachment

As additional validation, we performed simulations of charged droplet detachment for acetone, to ensure that the numerical model can be used to simulate generation of charged droplets. The relaxation time in acetone is in the order of 10 μs, so that interrupting the voltage pulse shortly before detachment is sufficient to ensure uncharged droplet generation. We therefore consider numerically the case of a voltage pulse interrupted at $t = 26$ ms, after the detachment of the droplet at $t = 21.8$ ms. The total charge inside of the computational domain during the detachment process is shown in Fig. 10.5. After the end of the electric pulse, the charge in the pendant liquid returns back to the capillary, and is replaced by a charge induced by the electric fields originating from the free falling droplet. The charge acquired by the droplet, $q = 149$ pC, is conserved during the free fall.

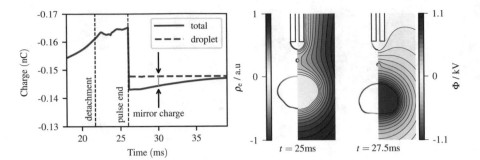

Fig. 10.5 Electric charge carried by the liquids during the detachment event. Left: total charge in the computational domain and in the droplet. Right: charge and potential distribution before and after droplet detachment

10.4 Conclusion

We have presented a model to solve electrohydrodynamic problems including dynamic charging of droplets. The electrical and mechanical problems are solved consistently on non moving grids using the Volume of Fluid interface capturing scheme, which ensure implicit handling of topology changes of the phase boundary. Simulated dynamics are compared with experimental dynamics and show a good agreement. The model can be readily applied to complex configurations involving multiple interacting droplets in electric fields.

Acknowledgements This work is funded by the German Research Foundation in the framework of the Collaborative Research Center Transregio 75.

References

1. Weckenmann, F., Bork, B., Oldenhof, E., Lamanna, G., Weigand, B., Boehm, B., Dreizler, A.: Single acetone droplets at supercritical pressure: droplet generation and characterization of PLIFP. Z. Phys. Chem.-Int. J. Res. Phys. Chem. Chem. Phys. **225**, 1417–1431 (2011)
2. Melcher, J.R., Taylor, G.I.: Electrohydrodynamics: a review of the role of interfacial shear stresses. Annu. Rev. Fluid Mech. **1**(1), 111–146 (1969)
3. Saville, D.A.: Electrohydrodynamics: the Taylor-Melcher leaky dielectric model. Annu. Rev. Fluid Mech. **29**(1), 27–64 (1997)
4. Openfoam. http://www.openfoam.org
5. Roisman, I.V., Opfer, L., Tropea, C., Raessi, M., Mostaghimi, J., Chandra, S.: Drop impact onto a dry surface: role of the dynamic contact angle. Colloids Surf. A Physicochem. Eng. Asp. **322**(1–3), 183–191 (2008)
6. Hirt, C.W., Nichols, B.D.: Volume of fluid (VOF) method for the dynamics of free boundaries. J. Comput. Phys. **39**(1), 201–225 (1981)

7. Tomar, G., Gerlach, D., Biswas, G., Alleborn, N., Sharma, A., Durst, F., Welch, S.W.J., Delgado, A.: Two-phase electrohydrodynamic simulations using a volume-of-fluid approach. J. Comput. Phys. **227**(2), 1267–1285 (2007)
8. Rohlfs, W., Dietze, G.F., Haustein, H.D., Kneer, R.: Two-phase electrohydrodynamic simulations using a volume-of-fluid approach: a comment. J. Comput. Phys. **231**(12), 4454–4463 (2012)

Part IV
Mathematical and Computational Methods Including Uncertainty Quantification

In this part of the book, we present papers that primarily focus on the development of new numerical methods for solving CEM problems including uncertainty quantification.

In their keynote talk on *"Multirate Shooting Method with Frequency Sweep for Circuit Simulation"*, K. Bittner and H.G. Brachtendorf introduce the multirate shooting technique, which is based on the reformulation of the system of differential-algebraic equations describing a circuit as partial differential equations first semi-discretized by Rothe's method in time and finally by the shooting method in space.

In *"A Trefftz Method for the Time-Harmonic Eddy Current Equation"*, R. Casagrande et al. present a new enriched finite element method that can efficiently deal with boundary layers and singularities typically arising along the conductor surface in eddy current problems under time-harmonic excitations. Inductive hardening is one important application.

The paper *"Survey on Semi-Explicit Time Integration of Eddy Current Problems"* by J. Dutiné et al. discusses highly efficient realizations of the explicit Euler method for eddy current problems. The finite element discretization of an eddy current problem results in a large system of differential-algebraic equations that can be reduced to an equivalent system of ordinary differential equations (ODE) by eliminating the unknowns in the non-conducting regions. This ODE system is then solved using the explicit Euler method.

In their paper *"A Local Mesh Modification Strategy for Interface Problems with Application to Shape and Topology Optimization"*, P. Gangl and U. Langer develop and analyze a local mesh modification technology that can follow a moving interface on a background triangular mesh without losing accuracy. The corresponding finite element discretization is used in the shape and topology optimization of electrical machines.

Q. Liu and R. Pulch provide a sensitivity analysis of linear dynamical systems in their contribution *"Numerical Methods for Derivative-Based Global Sensitivity Analysis in High Dimensions"*. They introduce numerical techniques for computing derivative-based sensitivity indices in the case of high-dimensional hypercubes.

These techniques are subsequently tested for a linear dynamical system that models a band-stop filter.

In *"Fitting Generalized Gaussian Distributions for Process Capability Index"*, T.G.J. Beelen et al. propose a new, fast and reliable numerical method for computing the defining parameters in generalized Gaussian density distributions.

In their paper *"Robust Optimization of an RFIC Isolation Problem under Uncertainties"*, P. Putek et al. incorporate uncertainty quantification into the modeling of electronic devices to provide reliable and robust simulation and optimization. The simulation is based on stochastic differential-algebraic equations. The new robust optimization methodology is then successfully applied to a Radio Frequency Integrated Circuit (RFIC) isolation problem.

Chapter 11
Multirate Shooting Method with Frequency Sweep for Circuit Simulation

Kai Bittner and Hans Georg Brachtendorf

Abstract We introduce multirate shootings methods to compute the response of radio frequency (RF) circuits with frequency modulated stimuli. The multirate technique is based on reformulating the system of ordinary differential algebraic equations (DAE) by partial differential equations (PDE). The PDE is semi-discretized by Rothe's method, i.e. by first discretizing the initial value problem. The resulting periodic boundary value problems are then solved by shooting techniques. Second, the instantaneous frequency is an additional unknown and concurrently estimated.

11.1 Introduction

The multirate simulation technique (see e.g. [1–4]) has been introduced to circuit simulation to handle RF signals with widely separated time scales in an efficient way, by reformulating the circuit equations as PDEs in different time scales. A semi-discretization of the multirate PDEs leads to a series of Periodic Steady State (PSS) problems, which are usually solved by waveform relaxation methods (Harmonic Balance, Finite Difference, Galerkin discretization), which approximate the periodic solution over a whole period. However, for some problems this can lead to convergence problems of nonlinear solvers (e.g. Newton) and large problem sizes with prohibitive memory and time requirements.

Here, we consider shooting (e.g. [5–11]) as an alternative approach for PSS. The advantage of shooting methods is that they can handle most problems for which a transient analysis is feasible. Furthermore, the size of linear and nonlinear problems to be solved is determined by the size of the circuit and not by the waveform, which can reduce memory requirements essentially and a speedup might be possible at least for some problems. Our goal here is to give a complete

K. Bittner (✉) · H. G. Brachtendorf
University of Applied Sciences of Upper Austria, Hagenberg, Austria
e-mail: Kai.Bittner@fh-hagenberg.at; Hans-Georg.Brachtendorf@fh-hagenberg.at

© Springer International Publishing AG, part of Springer Nature 2018
U. Langer et al. (eds.), *Scientific Computing in Electrical Engineering*,
Mathematics in Industry 28, https://doi.org/10.1007/978-3-319-75538-0_11

113

description of the shooting method in circuit simulation as an alternative to other PSS solvers.

We develop a shooting method based on multistep methods for circuit simulation in Sects. 11.2 and 11.3. Then shooting is applied to the multirate method in Sect. 11.4. Various approaches for an optimal frequency sweep are presented in Sect. 11.5. A numerical test in Sect. 11.6 concludes the results.

11.2   Circuit Equations and Multistep Methods

Consider the circuit equations in the charge/flux oriented modified nodal analysis (MNA) formulation, which yields a mathematical model in the form of a system of differential-algebraic equations (DAEs):

$$\frac{d}{dt} q\big(x(t)\big) + \underbrace{i\big(x(t)\big) + s(t)}_{g(x,t)} = 0, \qquad x(0) = x_0 \tag{11.1}$$

Here $x(t) \in \mathbb{R}^n$ is the vector of node potentials and specific branch currents and $q(x) \in \mathbb{R}^n$ is the vector of charges and fluxes. The vector $i(x) \in \mathbb{R}^n$ comprises static contributions, while $s(t) \in \mathbb{R}^n$ contains the contributions of independent sources.

The DAEs in (11.1) are usually solved by time integration formulas for stiff systems. Here, we consider implicit linear multistep methods, which approximate the solution at a discrete time step t_k based on approximations at previous time steps $t_\ell < t_k$ as follows. Let the approximations $x_\ell \approx x(t_\ell)$, $0 \le \ell < k$, be already computed. The approximation $x_k \approx x(t_k)$ is found as the solution of the nonlinear system

$$F(x_k) := \sum_{\ell=0}^{s_k} \alpha_\ell^k \, q(x_{k-\ell}) + \beta_\ell^k \, g(x_{k-\ell}, t_{k-\ell}) = 0, \qquad s_k \le k. \tag{11.2}$$

Usually the trapezoidal rule or Gear's backward difference formulas (BDF) are used in circuit simulation. A further choice, in particular for high Q oscillators, are the trigonometric BDF formulas from [12], which avoid artificial energy loss by numerical damping.

The nonlinear system is solved by Newton's method, where we need the Jacobian

$$DF(x_k) = \alpha_0^k \, C(x_k) + \beta_0^k \, G(x_k).$$

The computation of the Jacobians $C(x) := q'(x)$ and $G(x) := i'(x)$ is usually implemented in a circuit simulator, together with the evaluation of $q(x)$ and $g(x, t)$.

11.3 Periodic Steady States and Shooting

To determine the periodic steady state (PSS) of a circuit means to solve the periodic boundary value problem

$$\frac{d}{dt}q(x(t)) + i(x(t)) + s(t) = 0, \qquad x(0) = x(P) \tag{11.3}$$

instead of the initial value problem (11.1), where the source term $s(t)$ is required to be P-periodic to assure existence of a periodic solution. One approach to solve a boundary value problem are shooting methods [5–9]. The principal idea is to solve initial value problems and to adapt the initial values so that the boundary conditions are fulfilled. That is, a nonlinear equation for the boundary value problem has to be solved. However, the computation of the corresponding Jacobian matrix is not trivial in general. An approach for circuit simulation based on the backward Euler method was given in [10]. A generalization to BDF and trapezoidal rule can be found in [11]. Here we give a short description of the shooting method for periodic steady states of circuits, using multistep methods as introduced in Sect. 11.2.

In the sequel, we assume that a periodic steady state exists, and that the solution depends smoothly on the consistent initial value, which is true for many practical problems. Let $x(t; \xi)$ be the solution of the initial value problem with initial value ξ, i.e.,

$$\frac{d}{dt}q(x(t; \xi)) + i(x(t; \xi)) + s(t) = 0, \qquad x(0; \xi) = \xi. \tag{11.4}$$

Further, $\Phi(\xi) := x(P; \xi)$ is the value of the solution after one period. To find a PSS one has to determine an initial value $\xi \in \mathbb{R}^n$ such that $\Phi(\xi) - \xi = 0$, i.e., we have to solve a nonlinear system. For the application of Newton's method we need the Jacobian $\frac{d}{d\xi}(\Phi(\xi) - \xi) = \Phi'(\xi) - I$. Numerical differentiation is prohibitive expensive for larger circuits. Therefore, we consider an alternative approach.

Since $\Phi(\xi) = x(P; \xi)$ cannot be determined exactly, we replace $\Phi(\xi)$ by the approximation

$$\widetilde{\Phi}(\xi) := x_N := x_N(\xi),$$

where $x_k := x_k(\xi)$ is the solution from the multistep method (11.2) with $x_0 = \xi$ and $t_N = P$. While we need a consistent initial value ξ in (11.4), we can avoid this requirement if we use one or more (depending on the index of the DAE) backward Euler steps (BDF1) at the begin of the time integration [11, Lemma 4.2]. However, projecting the initial guess for Newton's method onto a consistent solution can improve the convergence of the shooting method [13].

The Jacobian $\widetilde{\Phi}'(\xi)$ is determined as follows. By differentiating (11.2) one obtains

$$\sum_{\ell=0}^{s_k} \left(\alpha_\ell^k C(x_{k-\ell}) + \beta_\ell^k G(x_{k-\ell})\right) \frac{dx_{k-\ell}}{d\xi} = 0. \tag{11.5}$$

This leads to the recursion

$$\frac{dx_k}{d\xi} = -\left(\alpha_0^k\, C(x_k) + \beta_0^k\, G(x_k)\right)^{-1}\left(\sum_{\ell=1}^{s_k}\left(\alpha_\ell^k\, C(x_{k-\ell}) + \beta_\ell^k G(x_{k-\ell})\right)\frac{dx_{k-\ell}}{d\xi}\right),$$

(11.6)

for $k = 1, \ldots, N$ with $\frac{dx_0}{d\xi} = I$.

Computing $\widetilde{\Phi}'(\xi) = \frac{dx_N}{d\xi}$ using direct solvers will be rather expensive. The good news are that $C(x_k)$ and $G(x_k)$ are sparse and that the LU-decomposition of the sparse matrix $\alpha_0^k\, C(x_k) + \beta_0^k\, G(x_k)$ has to be computed anyway in order to solve (11.2).[1] The bad news are that the inverse matrices are dense and thus the matrices $\frac{dx_k}{d\xi}$ are dense, too. That is, the computational complexity for computing $\widetilde{\Phi}'(\xi)$ amounts to $\mathcal{O}(n^{1+\gamma} N)$, if we assume the computational cost for the sparse forward-backward-substitutions (during Transient analysis) to be $\mathcal{O}(n^\gamma)$ for some $\gamma > 1$. Additionally, we have to solve a linear system with the dense matrix $\widetilde{\Phi}'(\xi) - I$, for the outer Newton iteration to determine ξ, which requires $\mathcal{O}(n^3)$ operations. However, this direct computation needs only limited memory of order $\mathcal{O}(n^2)$ (independent of N), if it is done immediately during the transient analysis.

However, we can attempt to solve the linear system by an iterative method, e.g., GMRES. This requires an efficient matrix vector multiplication $\widetilde{\Phi}'(\xi)\, y$ for any given $y \in \mathbb{R}^n$. Using the recursion (11.6) we can do this, without knowing the matrix $\widetilde{\Phi}'(\xi)$ itself. We define $y_k := \frac{dx_k}{d\xi}\, y$, which yields $y_N = \widetilde{\Phi}'(\xi)\, y$ by the recursion

$$\left(\alpha_0^k\, C(x_k) + \beta_0^k\, G(x_k)\right)y_k = \sum_{\ell=1}^{s}\left(\alpha_\ell^k\, C(x_{k-\ell}) + \beta_\ell^k G(x_{k-\ell})\right)y_{k-\ell}, \qquad y_0 = y.$$

(11.7)

This approach requires to store all matrices $C(x_k)$ and $G(x_k)$, $k = 0, \ldots, N$, since the vector $y = \widetilde{\Phi}(\xi) - \xi$ is only available after the transient analysis is complete. This results in a memory consumption $\mathcal{O}(N\, n^\gamma)$ for some small $\gamma > 1$. The computational cost of one matrix vector multiplication would be essentially the same as for the transient analysis (without device evaluation), i.e., $\mathcal{O}(N\, n^\gamma)$. This has to be multiplied by the number of iterations K needed by the iterative solver.

The iteration count K can be expected to be small in many cases due to the following statement (see e.g. [14, Prop. 4]). We assume that the matrix $\widetilde{\Phi}'(\xi)$ is diagonalizable, which is the typical case. The error for the residual r_m after m iteration steps can be estimated as

$$\|r_m\| \le c\, \|r_0\| \min_{p\in\Pi_{m-1},\, p(0)=1}\ \max_{\lambda\in\sigma}|p(\lambda)|$$

---

[1]We assume that a direct sparse solver is used to solve (11.2) and (11.6), which is reasonable in circuit simulation. For an iterative solver we have only to determine a preconditioner once for the multiple solves, i.e., similar considerations apply.

where c is a constant depending on the matrix, Π_{m-1} are the polynomials of degree less than m and σ is the set of eigenvalues of the system matrix $\tilde{\Phi}'(\xi)$. If the eigenvalues are clustered around few values one can choose polynomials with the zeros in this clusters to prove a fast decay of the residuals.

In a circuit many components of an initial value are damped out over a period of a periodic signal, which corresponds to small eigenvalues of $\tilde{\Phi}'(\xi)$, while only few eigenvalues may be away from zero (e.g. due to an oscillator). Thus, the Jacobian $\tilde{\Phi}'(\xi) - I$ may have mainly eigenvalues close to -1 with only a few exceptions, resulting in a fast convergence of GMRES, even without preconditioning.

The direct solver is suitable if n is not too large and accuracy of the linear solver is important, while for larger circuits the iterative solver might be favored.

11.4   Multirate Shooting Method

To separate different time scales, the circuit equation (11.1) can be replaced by partial differential equations [1–4], namely

$$\frac{\partial}{\partial \tau} q\big(\hat{x}(\tau, t)\big) + \omega(\tau) \frac{\partial}{\partial t} q\big(\hat{x}(\tau, t)\big) + i\big(\hat{x}(\tau, t)\big) + \hat{s}\big(\tau, t\big) = 0 \qquad (11.8)$$

where $\omega(\tau)$ is an estimate of the (scaled) angular frequency. The bivariate function $\hat{x}(\tau, t)$ is related to the univariate solution $x(t)$ of (11.1) as follows. For any solution $\hat{x}(\tau, t)$ of (11.8) we get by $x_\theta(t) = \hat{x}\big(t, \Omega_\theta(t)\big)$, $\Omega_\theta(t) = \theta + \int_0^t \omega(s)\, ds$ a solution of

$$\frac{d}{dt} q\big(x(t)\big) + i\big(x(t)\big) + \hat{s}\big(t, \Omega_\theta(t)\big) = 0.$$

Thus, if we choose $\hat{s}$ such that $s(t) = \hat{s}\big(t, \Omega_0(t)\big)$, then the solution of (11.8) provides also a solution of (11.1), i.e., $x(t) = x_0(t) = \hat{x}\big(t, \Omega_0(t)\big)$.

The multirate equation (11.8) are usually solved under periodicity conditions $\hat{x}(\tau, t) = \hat{x}(\tau, t + P)$ in t and initial conditions $\hat{x}(0, t) = X_0(t)$ in τ. The source term has then to be periodic, too, i.e., $\hat{s}(\tau, t) = \hat{s}(\tau, t + P)$. The term $\omega(\tau)$ can be used to adapt to frequency modulated signals. In [3] it was shown that we can improve the smoothness of $\hat{x}$ in τ, if P and $\omega(\tau)$ are chosen such that $\frac{\omega(\tau)}{P}$ equals the instantaneous frequency.

Following [3, 4] we use Rothe's method for semi-discretization. Using Gear's BDF^2 method of order s with respect to τ one obtains

$$\sum_{i=0}^{s} \tilde{\alpha}_i^k q\big(X_{k-i}(t)\big) + \omega_k \frac{d}{dt} q\big(X_k(t)\big) + i\big(X_k(t)\big) + \hat{s}\big(\tau_k, t\big) = 0 \qquad (11.9)$$

---

^{2}Other multistep method (e.g. trapezoidal rule) can be used, too.

With the definition

$$g_k(x, t) := \tilde{\alpha}_0^k q(x) + i(x) + \hat{s}(\tau_k, t) + \sum_{i=1}^{s} \tilde{\alpha}_i^k q\big(X_{k-i}(t)\big),$$

X_k is the solution of the periodic boundary value problem

$$\omega_k \frac{d}{dt} q\big(x(t)\big) + g_k(x(t), t) = 0, \qquad x(t) = x(t + P). \qquad (11.10)$$

The new problem (11.10) is closely related to the original periodic steady state problem of the circuit, only modified by the additional 'source term' $\sum_{i=1}^{s} \tilde{\alpha}_i^k q\big(X_{k-i}(t)\big)$. Analogous to (11.5) one obtains

$$\sum_{\ell=0}^{s_m} \Big(\omega_k \alpha_\ell^m \, C(x_{m-\ell}) + \beta_\ell^m \big(G(x_{m-\ell}) + \tilde{\alpha}_0^k \, C(x_{m-\ell})\big)\Big) \frac{dx_{m-\ell}}{d\xi} = 0. \qquad (11.11)$$

Thus, X_k can be approximated by the shooting method from Sect. 11.3. The only additional problem is to compute $\sum_{i=1}^{s} \tilde{\alpha}_i^k q\big(X_{k-i}(t)\big)$ at the transient time steps for $t_{k,i}$ for X_k. This requires to store the values $q_\ell = q(X_{\ell,i}) \approx q(X_\ell(t_{\ell,i}))$. These values can be used to approximate $q\big(X_\ell(t_{k,i})\big), k > \ell$, e.g. by interpolation.

11.5   Frequency Sweep and Smoothness Conditions

As pointed out in [3, 15, 16] the function $\omega(\tau)$ can be chosen in order to obtain a smoother solution, which accelerates the simulation due to larger time steps in τ. The observation that a modification of $\omega(\tau)$ for $\tau < \tau_k$ results in a phase shift of $\hat{x}(\tau_k, \cdot)$ leads to the proposition of a smoothness condition of the form

$$\big\| \tfrac{\partial}{\partial \tau} \hat{x}(\tau, \cdot) \big\|_{L^2} \to \min, \qquad (11.12)$$

(cf. [3, 15, 16]) or

$$\big\| \tfrac{\partial}{\partial \tau} q\big(\hat{x}(\tau, \cdot)\big) \big\|_{L^2} \to \min, \qquad (11.13)$$

(cf. [15–17]) which should reduce changes with respect to τ. Here, we use the norm $\|x\|_{L^2}^2 := \int_0^P \sum_{k=1}^n |x_k(t)|^2 \, dt$ and the corresponding inner product $\langle x, y \rangle := \int_0^P \sum_{k=1}^n x_k(t) \, y_k(t) \, dt$. In many cases a (near) optimal choice of $\omega(\tau)$ is known in advance, e.g., from the (instantaneous) frequency of sources in a driven circuit.

Often, a good choice of $\omega(\tau)$ is not known in advance, but central for the success of the simulation, e.g., for the simulation of high Q oscillators without numerical damping [12] or voltage controlled oscillators (VCO) (in a Phase-Locked

Loop (PLL)) [3, 4]. For the existence and uniqueness of an optimal $\omega(\tau)$ we refer to [18]. There have been several approaches to include the above smoothness conditions into simulation methods, e.g. if finite differences or collocation or Galerkin methods (Harmonic Balance, spline wavelets) are applied to solve the periodic problem (11.10). During our investigations it turned out that the application of these approaches to the shooting method is not straightforward. In this section we will develop frequency sweep following methods for the shooting methods, based on established methods for other periodic solvers.

11.5.1 An Explicit Approach

We first refer to an approach of Houben [17] based on condition (11.13), which leads to the equality

$$\omega(\tau) = -\frac{\left\langle \frac{\partial}{\partial t} q\big(\hat{x}(\tau, \cdot)\big), i\big(\hat{x}(\tau, \cdot)\big) + \hat{s}\big(\tau, \cdot\big) \right\rangle}{\left\| \frac{\partial}{\partial t} q\big(\hat{x}(\tau, \cdot)\big) \right\|_{L^2}^2}. \tag{11.14}$$

Apparently, we can determine $\omega(\tau)$ only after $\hat{x}(\tau, \cdot)$ is known. However in the Rothe discretization (11.9) we can use the guess

$$\omega_k \approx \omega(\tau_{k-1}) = -\frac{\left\langle \frac{d}{dt} q(X_{k-1}), i(X_{k-1}) + \hat{s}\big(\tau_{k-1}, \cdot\big) \right\rangle}{\left\| \frac{d}{dt} q(X_{k-1}) \right\|_{L^2}^2}, \tag{11.15}$$

based on the solution of the previous time step. From the shooting method we know not only approximations $X_{k-1,\ell} \approx X_{k-1}(t_\ell)$, but also $g_{k-1,\ell} = i\big(X_{k-1}(t_\ell)\big) + \hat{s}\big(\tau_{k-1}, t_\ell\big)$ and $q_{k-1,\ell} = q\big(X_{k-1}(t_\ell)\big)$. Approximations $Dq_{k-1,\ell} = \frac{d}{dt} q\big(X_{k-1}(t_\ell)\big)$ for the derivatives can be computed using finite differences, as it is done in the BDF method anyway. Numerical integration leads to the formulation

$$\omega_k = -\frac{\sum_{\ell=0}^N w_{k,\ell} \langle Dq_{k-1,\ell}, g_{k-1,\ell} \rangle}{\sum_{\ell=0}^N w_{k,\ell} \| Dq_{k-1,\ell} \|_{L^2}^2}, \tag{11.16}$$

where the $w_{k\ell}$ are quadrature weights for the grid $\{t_{k,\ell}\}$. The original approach in [17] uses the method of lines, but it works with Rothe's method as well.

Although the method is simple and easy to implement, also in a shooting method, it has limits. Since the computation uses only data from the previous time step, the accuracy of this approach is limited. In many circuits an accurate estimate for the optimal ω_k is essential for the efficiency of the multirate algorithm. In the sequel we will consider methods, where ω_k is determined in the Newton iteration of the shooting method for X_k.

11.5.2 An Additional Equation

Treating ω_k in Eq. (11.10) as an unknown requires two things. We need derivatives with respect to ω_k for the Jacobian, and an additional equation and its derivatives. Since our shooting equation depends now also on ω, we replace $\widetilde{\Phi}(\xi)$ by $\widetilde{\Phi}(\xi, \omega)$. We obtain the derivative with respect to ω as $\frac{\partial}{\partial \omega}\left(\widetilde{\Phi}(\xi, \omega) - \xi\right) = \frac{\partial}{\partial \omega}\widetilde{\Phi}(\xi, \omega) = \frac{dx_N}{d\omega}$. Obviously the initial value is independent of ω, i.e., $\frac{dx_0}{d\omega} = 0$. Differentiating the discretized version of Eq. (11.10) with respect to ω yields

$$\sum_{\ell=0}^{s_m} \alpha_\ell^m \left(q\left(x_{m-\ell}\right) + \omega C(x_{m-\ell})\frac{dx_{m-\ell}}{d\omega}\right) + \beta_\ell^m \left(G_k(x_{m-\ell}) + \tilde{\alpha}_0^k C(x_{m-\ell})\right)\frac{dx_{m-\ell}}{d\omega} = 0.$$

(11.17)

Thus $\frac{\partial}{\partial \omega}\widetilde{\Phi}(\xi, \omega)$ can be computed during the transient simulation of the shooting using the recursion over $\frac{dx_k}{d\omega}$ similar to (11.6). Pulch [15, 16] suggests the following approach. Based on the Gâteaux derivative, he shows that the smoothness condition (11.12) is equivalent to

$$0 = \langle \tfrac{\partial}{\partial \tau}\hat{x}(\tau, \cdot), \tfrac{\partial}{\partial t}\hat{x}(\tau, \cdot)\rangle.$$

(11.18)

The challenge is to incorporate this equation into the shooting method, which is done as follows. First we semi-discretize by replacing $\frac{\partial}{\partial \tau}\hat{x}(\tau_k, \cdot)$ and $\frac{\partial}{\partial t}\hat{x}(\tau_k, \cdot)$ by $\frac{X_k - X_{k-1}}{\tau_k - \tau_{k-1}}$ and X_k', respectively. Thus condition (11.18) is substituted by

$$0 = \langle X_k - X_{k-1}, X_k'\rangle$$
$$= \tfrac{1}{2}\int_0^P \tfrac{d}{dt}\|X_k(t)\|^2 \, dt - \left(\underbrace{X_{k-1}(P)^T X_k(P) - X_{k-1}(0)^T X_k(0)}_{X_{k-1}(0)}\right) + \int_0^P X_{k-1}'(t)^T X_k(t) \, dt$$
$$= \tfrac{1}{2}\left(\|X_k(P)\|^2 - \|X_k(0)\|^2\right) + X_{k-1}(0)^T\left(X_k(0) - X_k(P)\right) + \int_0^P X_{k-1}'(t)^T X_k(t) \, dt,$$

i.e., the solution (x, ω_k) of the periodic problem (11.10) shall satisfy

$$\tfrac{1}{2}\left(\|x(P)\|^2 - \|x(0)\|^2\right) + X_{k-1}(0)^T\left(x(0) - x(P)\right) + \int_0^P X_{k-1}'(t)^T x(t) \, dt = 0.$$

(11.19)

Although x is P-periodic, we cannot assume $x(P) = x(0)$ during the Newton iteration of the shooting method, i.e., none of terms above can be neglected.

By numerical integration we approximate (11.19) by

$$\Psi(\xi, \omega) := \tfrac{1}{2}\left(x_N^T x_N - x_0^T x_0\right) + \tilde{x}_{k-1,0}^T\left(x_0 - x_N\right) + \sum_{i=0}^N w_i \, \tilde{x}_{k-1,i}^T x_i = 0,$$

(11.20)

where the w_k are quadrature weights and $\tilde{x}_{k-1,i}$ are approximations of $X'_{k-1}(t_{k,i})$. Now we have to solve the system $\widetilde{\Phi}(\xi, \omega) = 0$, $\Psi(\xi, \omega) = 0$ with $n + 1$ unknowns and $n + 1$ equations.

For Newton's method one needs the derivatives of $\Psi(\xi, \omega)$, which are

$$\tfrac{\partial}{\partial \xi} \Psi(\xi, \omega) = \left(x_N - \tilde{x}_{k-1,0}\right)^T \frac{dx_N}{d\xi} - \left(x_0 - \tilde{x}_{k-1,0}\right)^T + \sum_{i=0}^{N} w_i\, \tilde{x}_{k-1,i}^T \frac{dx_i}{d\xi},$$

$$\tfrac{\partial}{\partial \omega} \Psi(\xi, \omega) = \left(x_N - \tilde{x}_{k-1,0}\right)^T \frac{dx_N}{d\omega} + \sum_{i=0}^{N} w_i\, \tilde{x}_{k-1,i}^T \frac{dx_i}{d\omega}.$$

This does not increase the computational cost essentially, since $\frac{dx_i}{d\xi}$ is already computed during the recursion (11.6) (with the modification from (11.11)) and $\frac{dx_i}{d\omega}$ is determined in the recursion (11.17). The only extra effort is to add up the terms $\frac{dx_i}{d\xi} X'_{k-1}(t_{k,i})$ and $\frac{dx_i}{d\omega} X'_{k-1}(t_{k,i})$ during the computation.

If we replace (11.12) by (11.13) one obtains by an analogous argument

$$\Psi_q(\xi, \omega) := \tfrac{1}{2}\left(q_N^T q_N - q_0^T q_0\right) + \tilde{q}_{k-1,0}^T\left(q_0 - q_N\right) + \sum_{i=0}^{N} w_i \tilde{q}_i^T q_i = 0, \quad (11.21)$$

as well as the derivatives with respect to ξ and ω.

11.5.3 A Discrete Smoothness Criterion

We start from the smoothness criterion (11.12), which we discretize instead of formulating an equivalent equation, namely as

$$\|X_k(t) - X_{k-1}(t)\|_{L^2}^2 \to \min. \quad (11.22)$$

A similar criterion was introduced in [3] for waveform relaxation methods. Using numerical integration, condition (11.22) becomes

$$\sum_{\ell=0}^{N} w_\ell \left|x_\ell(\xi, \omega) - \tilde{x}_{k-1,\ell}\right|^2 \to \min, \quad (11.23)$$

with suitable quadrature weights w_ℓ and approximations $\tilde{x}_{k-1,\ell}$ of $X_{k-1}(t_{k,\ell})$. This optimization condition has to be solved under the condition that $\Phi(\xi, \omega) - \xi = 0$ (remember $x_0(\xi, \omega) = \xi$ and $x_N(\xi, \omega) = \Phi(\xi, \omega)$ with $t_N = P$). Using a Lagrange multiplier approach we obtain

$$\tfrac{1}{2} \sum_{\ell=0}^{N} w_\ell \left| x_\ell(\xi, \omega) - \tilde{x}_{k-1,\ell} \right|^2 + \lambda^T \left(\Phi(\xi, \omega) - \xi \right) \to \min. \tag{11.24}$$

To establish a Gauss-Newton type method, we linearize the problem as follows. For a given initial guess (ξ, ω) we use the linear approximation

$$x_\ell(\xi - d_\xi, \omega - d_\omega) \approx x_\ell(\xi, \omega) - \tfrac{\partial x_\ell}{\partial \xi}(\xi, \omega)\, d_\xi - d_\omega\, \tfrac{\partial x_\ell}{\partial \omega}(\xi, \omega)$$

$$\Phi(\xi - d_\xi, \omega - d_\omega) - (\xi - d_\xi)$$
$$\approx \Phi(\xi, \omega) - \xi - \left(\tfrac{\partial \Phi}{\partial \xi}(\xi, \omega) - I \right) d_\xi - d_\omega\, \tfrac{\partial \Phi}{\partial \omega}(\xi, \omega).$$

Substituting this into (11.24) and setting the derivatives with respect to d_ξ, d_ω, and λ to zero we obtain the equations

$$-\sum_{\ell=0}^{N} w_\ell \left(\tfrac{\partial x_\ell}{\partial \xi}(\xi, \omega) \right)^T \left(x_\ell(\xi, \omega) - \tfrac{\partial x_\ell}{\partial \xi}(\xi, \omega)\, d_\xi - d_\omega\, \tfrac{\partial}{\partial \omega} x_\ell(\xi, \omega) - \tilde{x}_{k-1,\ell} \right)$$
$$-\left(\tfrac{\partial}{\partial \xi} \Phi(\xi, \omega) - I \right)^T \lambda = 0$$

$$-\sum_{\ell=0}^{N} w_\ell \left(\tfrac{\partial x_\ell}{\partial \omega}(\xi, \omega) \right)^T \left(x_\ell(\xi, \omega) - \tfrac{\partial x_\ell}{\partial \xi}(\xi, \omega)\, d_\xi - d_\omega\, \tfrac{\partial}{\partial \omega} x_\ell(\xi, \omega) - \tilde{x}_{k-1,\ell} \right)$$
$$-\left(\tfrac{\partial}{\partial \omega} \Phi(\xi, \omega) \right)^T \lambda = 0$$

$$\Phi(\xi, \omega) - \xi - \left(\tfrac{\partial \Phi}{\partial \xi}(\xi, \omega) - I \right) d_\xi - d_\omega\, \tfrac{\partial \Phi}{\partial \omega}(\xi, \omega) = 0.$$

For abbreviation we introduce $U, A \in \mathbb{R}^{N \times N}$, $v, c, z, b \in \mathbb{R}^N$ and $\rho, \eta \in \mathbb{R}$

$$U := \sum_{\ell=0}^{N} w_\ell \left(\tfrac{\partial x_\ell}{\partial \xi}(\xi, \omega) \right)^T \tfrac{\partial x_\ell}{\partial \xi}(\xi, \omega), \quad A := \tfrac{\partial \Phi}{\partial \xi}(\xi, \omega) - I; \tag{11.25}$$

$$v := \sum_{\ell=0}^{N} w_\ell \left(\tfrac{\partial x_\ell}{\partial \xi}(\xi, \omega) \right)^T \tfrac{\partial x_\ell}{\partial \omega}(\xi, \omega), \quad c := \sum_{\ell=0}^{N} w_\ell \left(\tfrac{\partial x_\ell}{\partial \xi}(\xi, \omega) \right)^T \left(x_\ell(\xi, \omega) - \tilde{x}_{k-1,\ell} \right),$$

$$z := \tfrac{\partial \Phi}{\partial \omega}(\xi, \omega), \qquad\qquad b := \Phi(\xi, \omega) - \xi;$$

$$\rho := \sum_{\ell=0}^{N} w_\ell \left(\tfrac{\partial x_\ell}{\partial \omega}(\xi, \omega) \right)^T \tfrac{\partial x_\ell}{\partial \omega}(\xi, \omega), \quad \eta := \sum_{\ell=0}^{N} w_\ell \left(\tfrac{\partial x_\ell}{\partial \omega}(\xi, \omega) \right)^T \left(x_\ell(\xi, \omega) - \tilde{x}_{k-1,\ell} \right)$$

such that the above linear system becomes (in block matrix notation)

$$\begin{pmatrix} U & v & A^T \\ v^T & \rho & z^T \\ A & z & 0 \end{pmatrix} \cdot \begin{pmatrix} d_\xi \\ d_\omega \\ \lambda \end{pmatrix} = \begin{pmatrix} c \\ \eta \\ b \end{pmatrix}.$$

By a Schur complement elimination we obtain the solutions

$$d_\omega = \frac{\eta - v^T \tilde{b} - \tilde{z}^T \left(c - U\tilde{b}\right)}{\rho - 2v^T \tilde{z} + \tilde{z}^T U \tilde{z}} \qquad \text{and} \qquad d_\xi = \tilde{b} - d_\omega \tilde{z}.$$

for Newton updates of ξ and ω, where $A\tilde{b} = b$ and $A\tilde{z} = z$.

This requires to solve a linear system with two right hand sides, which can be done nearly as efficient as in the original shooting. However, the computation of U is rather expensive since it requires $N + 1$ matrix-matrix multiplications. A faster way is to compute first the vectors $\zeta_\ell = \frac{\partial x_\ell}{\partial \xi}(\xi, \omega)\,\tilde{z}$. Then $\tilde{z}^T U \tilde{z}$ is computed by

$$\tilde{z}^T U \tilde{z} = \sum_{\ell=0}^{N} w_\ell \zeta_\ell^T \zeta_\ell, \tag{11.26}$$

which needs only $N + 1$ inner products and $N + 1$ matrix vector products. The value of $\tilde{z}^T U \tilde{b}$ can be computed analogously. However, we need to store the Jacobians $\frac{\partial x_\ell}{\partial \xi}(\xi, \omega)$. That is, if we follow the memory saving approach (with a direct shooting solver) we will compute U directly using the formula in (11.25). For the time saving approach (with GMRES in shooting) we will use the stored data to do a fast computation based on (11.26) (cf. Sect. 11.3).

11.6 Numerical Test

The described methods have been implemented in C++ and incorporated in our circuit simulator LinzFrame. We have tested the method on a PLL (containing 145 MOSFETs and 80 unknowns), leading to a DAE of index 1. Here we show the multirate simulation of the locking phase using the frequency sweep method from Sect. 11.5.2. In Fig. 11.1 one can see that the reference and feedback signal are in-phase after ca. 200 μs. This is reflected by the charge pump output in Fig. 11.2 which measures the phase difference of both signals and is low pass filtered to control the VCO. The instantaneous frequency estimate in Fig. 11.2 provides information on the frequency modulation of the signals.

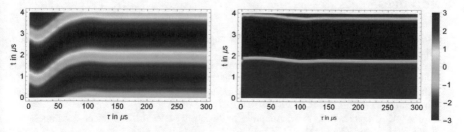

Fig. 11.1 Reference and feedback signal of the PLL

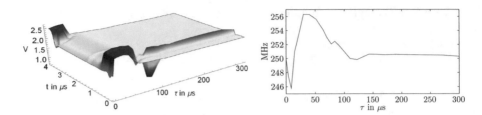

Fig. 11.2 Charge pump output and instantaneous frequency estimate based on $\omega(\tau)$

It turned out that the shooting method performs much better than e.g. the adaptive spline-Galerkin method from [4] in this locking phase. This is due to the fact that the transient simulation does not rely on a good initial guess for the PSS, which is taken from the previous envelope time step for Galerkin or finite difference schemes. In the locking phase, signals as the charge pump output depicted in Fig. 11.2 will require very small envelope time steps to achieve convergence of Newton's method. However, if all signals are sufficiently smooth (e.g. after locking of the PLL), adaptive Galerkin or FD schemes often perform better since they can employ information on grid and signal shape from the previous envelope time step.

11.7 Conclusion

A shooting method to determine PSS of circuits has been developed and implemented. Possible modifications of this method have been introduced to solve sub-problems in the PDE based multirate circuit simulation method for RF circuits. The new method provides an alternative to waveform relaxation methods if the latter fail due to prohibitive time or memory requirements, or convergence problems.

Acknowledgements This work has been partly supported by the fp7 project nanoCOPS under grant 619166 and the EFRE project Connected Vehicles under grant IWB2020.

References

1. Brachtendorf, H.G.: Simulation des eingeschwungenen Verhaltens elektronischer Schaltungen. Shaker, Aachen (1994)
2. Brachtendorf, H.G., Welsch, G., Laur, R., Bunse-Gerstner, A.: Numerical steady state analysis of electronic circuits driven by multi-tone signals. Electr. Eng. **79**(2), 103–112 (1996)
3. Bittner, K., Brachtendorf, H.G.: Optimal frequency sweep method in multi-rate circuit simulation. COMPEL **33**(4), 1189–1197 (2014)
4. Bittner, K., Brachtendorf, H.G.: Adaptive multi-rate wavelet method for circuit simulation. Radioengineering **23**(1), 300–307 (2014)

5. Aprille, T.J., Trick, T.N.: Steady-state analysis of nonlinear circuits with periodic inputs. Proc. IEEE **60**(1), 108–114 (1972)
6. Strohband, P.H., Laur, R., Engl, W.: TNPT - an efficient method to simulate forced RF networks in time domain. IEEE J. Solid-State circuits **12**(3), 243–246 (1977)
7. Skelboe, S.: Computation of the periodic steady-state response of nonlinear networks by extrapolation methods. IEEE Trans. Circuits Syst. **CAS-27**(3), 161–175 (1980)
8. Kakizaki, M., Sugawara, T.: A modified Newton method for the steady-state analysis. IEEE Trans. Comput. Aided Des. **4**(4), 662–667 (1985)
9. Stoer, J., Bulirsch, R.: Introduction to Numerical Analysis, 2nd edn. Springer, New York (1992)
10. Telichevesky, R., Kundert, K.S., White, J.K.: Efficient steady-state analysis based on matrix-free Krylov-subspace methods. In: Proceedings of the 32nd Design Automation Conference, DAC-95, pp. 480–484. ACM, New York (1995)
11. Baiz, A.: Effiziente Lösung periodischer differential-algebraischer Gleichungen in der Schaltungssimulation. Ph.D. thesis, TU Darmstadt (2003). Shaker Verlag, Aachen
12. Brachtendorf, H., Bittner, K.: Grid size adapted multistep methods for high Q oscillators. IEEE Trans. Comput. Aided Des. Integr. Circuits Syst. **32**(11), 1682–1693 (2013)
13. Brachtendorf, H.G., Laur, R.: On consistent initial conditions for circuit's DAEs with higher index. IEEE Trans. Circuits Syst. I, Fundam. Theory Appl. **48**(5), 606–612 (2001)
14. Saad, Y., Schultz, M.: GMRES a generalized minimal residual algorithm for solving nonsymmetric linear system. SIAM J. Sci. Stat. Comput. **7**, 856–869 (1986)
15. Pulch, R.: Initial-boundary value problems of warped MPDAEs including minimisation criteria. Math. Comput. Simul. **79**, 117–132 (2008)
16. Pulch, R.: Variational methods for solving warped multirate partial differential algebraic equations. SIAM J. Sci. Comput. **31**(2), 1016–1034 (2008)
17. Houben, S.: Simulating multi-tone free-running oscillators with optimal sweep following. In: Schilders, W., ter Maten, E., Houben, S. (eds.): Scientific Computing in Electrical Engineering 2002. Mathematics in Industry, pp. 240–247. Springer, Berlin (2004)
18. Kugelmann, B., Pulch, R.: Existence and uniqueness of optimal solutions for multirate partial differential algebraic equations. Appl. Numer. Math. **97**, 69–87 (2015)

Chapter 12
A Trefftz Method for the Time-Harmonic Eddy Current Equation

Check for updates

Raffael Casagrande, Christoph Winkelmann, Ralf Hiptmair, and Jörg Ostrowski

Abstract We present a discontinuous finite element method to resolve the skin effect in conductors on coarse meshes. The idea is to take into account the exponential decay in the finite element trial space, which enables to resolve the skin layer independent of the size of the mesh cells. The discontinuous, Trefftz-type basis functions are coupled across the element boundaries by the interior penalty-/Nitsche's method and numerical experiments affirm the effectiveness of the method for thin boundary layers.

12.1 Introduction

We consider the vector potential formulation of the eddy current problem in the frequency domain with temporal gauge ($\varphi = 0$),

$$\mathbf{curl}\left(\mu^{-1}\mathbf{curlA}\right) + i\omega\sigma\mathbf{A} = \mathbf{j}^i. \tag{12.1}$$

Here

- $\mathbf{A}(\mathbf{x})$ is a vector potential,
- $\mathbf{B} = \nabla \times \mathbf{A}$ is the magnetic flux density,
- $\mathbf{j}^i(\mathbf{x})$ is the impressed, solenoidal electric current,
- $\omega > 0$ is the angular frequency, and
- $\sigma(\mathbf{x})$ is the electric conductivity (which can be zero in parts of the domain and is assumed to be piecewise constant).

R. Casagrande (✉) · C. Winkelmann · R. Hiptmair
Seminar for Applied Mathematics, ETH Zürich, Zürich, Switzerland
e-mail: raffael.casagrande@sam.math.ethz.ch; christoph.winkelmann@sam.math.ethz.ch;
hiptmair@sam.math.ethz.ch

J. Ostrowski
ABB Switzerland Ltd., Corporate Research, Baden, Switzerland
e-mail: joerg.ostrowski@ch.abb.com

© Springer International Publishing AG, part of Springer Nature 2018
U. Langer et al. (eds.), *Scientific Computing in Electrical Engineering*,
Mathematics in Industry 28, https://doi.org/10.1007/978-3-319-75538-0_12

127

It is well known that the solution of (12.1) exhibits singularities in edges (and corners) of conductors [4], as well as exponential boundary layers along the surface of conductors (*skin effect*). I.e. the induced current $i\omega\sigma\mathbf{A}$ is concentrated at the surface of conductors and decays rapidly towards the interior. The thickness of the boundary layer is characterized by the skin-depth δ.

Induction has many applications in industry. An example is inductive hardening [8], where the workpiece is heated quickly at the surface, and is then rapidly cooled down before the heat is distributed into the interior by heat conduction. In this case the skin effect plays a fundamental technical role and resolving the skin layer is essential.

For the classical low order Finite Element Method (FEM) this means that the boundary layers must be resolved by the underlying mesh. This can be achieved by adapting the mesh manually or by refining an existing mesh towards the boundary layers, which can be automated (h-refinement). However, in industrial applications the skin depth δ can be orders of magnitude smaller than the diameter of the conductor so that the mesh must be refined multiple times towards the boundary layer(s). This leads to a vast increase in the number of degrees of freedom (DOF) which may render the solution of the linear system prohibitively expensive.

Alternatively one can refine the mesh just once to create a mesh layer of thickness $O(k\delta)$ where k is the polynomial degree of the test functions [9, 10]. However creating such a 3D mesh for industrial applications can be hard, especially if tetrahedral elements are used.

A partial remedy for this problem are Impedance Boundary Conditions (IBC) [8]: The conductor is replaced by Robin-type boundary conditions and the electromagnetic fields are only calculated at the surface of the conductor. Since the IBC approximation assumes that the conductor surface is flat, the solution deteriorates as the radius of curvature of the conductor surface becomes comparable to the skin-depth δ. In particular the IBC solution deviates strongly from the physically correct fields at edges and corners of the conductor.

In this work we propose to resolve the boundary layers directly on coarse meshes (we assume the meshsize $h \gg \delta$) by enriching the approximation space with suitable functions. More precisely, our approximation space will contain two types of (discontinuous) basis functions:

- Edge elements R_k [7], and
- Exponential boundary layer functions modulated/multiplied with polynomials.

We deal with the discontinuous nature of the basis functions in the framework of Discontinuous Galerkin (DG) methods and discretize (12.1) by the Non-Symmetric Weighted Interior Penalty (NWIP) method [3].

12.2 Non-symmetric Weighted Interior Penalty Framework

We consider the time-harmonic eddy current equation (12.1) on a bounded, open, polyhedral domain $\Omega \subset \mathbb{R}^3$ with Lipschitz boundary. Furthermore we denote by $\Omega_0 \subset \Omega$ the open subdomain where $\sigma = 0$ and define $\Omega_\sigma = \Omega \setminus \overline{\Omega_0}$.

Perturbed Problem It is well-known that the time-harmonic eddy current equation (12.1) does not uniquely determine the vector potential $\mathbf{A}$ in Ω_0, i.e. (12.1) is an ungauged formulation. In this work we restore the uniqueness of $\mathbf{A}$ by considering the *perturbed* time-harmonic eddy current problem [1],

$$\mathbf{curl}\left(\mu^{-1}\mathbf{curl}\mathbf{A}^\alpha\right) + \kappa^\alpha \mathbf{A}^\alpha = \mathbf{j}^i, \qquad \text{in } L^2(\Omega)^3 \qquad (12.2a)$$

$$\mathbf{n} \times \mathbf{A}^\alpha = 0 \qquad \text{on } \partial\Omega. \qquad (12.2b)$$

Here the boundary condition (12.2b) implies $\mathbf{n} \cdot \mathbf{curl}\mathbf{A} = \mathbf{n} \cdot \mathbf{B} = 0$ which reflects the decay of the magnetic field far away from the source $\mathbf{j}^i$. Moreover,

$$\kappa^\alpha(\mathbf{x}) := \begin{cases} i\omega\sigma(\mathbf{x}) & \text{for } x \in \Omega_\sigma, \\ \alpha & \text{for } x \in \Omega_0, \end{cases}$$

with $\alpha > 0$ being the regularization parameter. One expects that for $\alpha \to 0$ also $\mathbf{A}^\alpha \to \mathbf{A}$, or more precisely [1, Lemma 33],

Lemma 12.1 *Under the above assumptions we have,* $\|\mathbf{A} - \mathbf{A}^\alpha\|_{\mathbf{H}(\mathbf{curl};\Omega)} \leq C\alpha \|\mathbf{A}\|_{L^2(\Omega)^3}$, *where C is independent of α but depends on μ, σ, ω and the domain Ω.*

Broken Sobolev Spaces We assume that there exists a partition $P_\Omega = \{\Omega_i\}_i$ such that each Ω_i is a polyhedron and such that the permeability $0 < \mu < \infty$ and the coefficient function $0 < \kappa^\alpha < \infty$ are constant on each Ω_i. We will assume that the solution $\mathbf{A}^\alpha$ lies in the broken Sobolev space

$$V^*(P_\Omega) := \left\{ \mathbf{A} \in L^2(\Omega)^3 \,\middle|\, \mathbf{A}|_K \in H^1(K)^3, \ \mathbf{curl}\mathbf{A}|_K \in H^1(K)^3 \ \forall K \in P_\Omega \right\}.$$

Here $H^1(K) := \{f \in L^2(K)| \mathbf{grad}\, f \in L^2(K)^3\}$ denotes the usual Sobolev space.

Meshes, Jumps, Averages Let $\mathscr{T}_h$ denote a hybrid (tetrahedras, pyramids, prisms, hexahedras), affine, conforming mesh on Ω that is *compatible* with the partition P_Ω, that is every mesh element $T \in \mathscr{T}_h$ lies in exactly one $\Omega_i \in P_\Omega$. Thus κ^α, μ are constant on every mesh cell $T \in \mathscr{T}_h$ and we have $V^*(P_\Omega) \subset V^*(\mathscr{T}_h)$. Furthermore we let $\mathscr{F}_h^i$ denote the set of inner intersections of $\mathscr{T}_h$ and define the tangential jump and *weighted* average of a vector valued function $\mathbf{A} \in V^*(\mathscr{T}_h)$ on an inner face $F \in \mathscr{F}_h^i$, $F = \partial T_i \cap \partial T_j$, as follows:

$$[\![\mathbf{A}_h]\!]_T = \mathbf{n}_F \times \left(\mathbf{A}_h|_{T_i} - \mathbf{A}_h|_{T_j}\right), \qquad \text{(jump)}$$

$$\{\!\!\{\mathbf{A}_h\}\!\!\}_w = w_1 \mathbf{A}_h|_{T_i} + w_2 \mathbf{A}_h|_{T_j}, \qquad \text{(average)}$$

Here $\mathbf{n}_F$ always points from T_i to T_j and $w_i \in [0, 1]$ are such that $w_1 + w_2 = 1$.

NWIP-Formulation We discretize the perturbed eddy current problem (12.2) using a *finite dimensional* subspace $V_h \subset V_h^* := \{\mathbf{A} \in V^*(\mathscr{T}_h) \mid \mathbf{n} \times \mathbf{A} = 0 \text{ on } \partial\Omega\}$. Multiplying (12.2) with a discontinuous test function $\mathbf{A}_h' \in V_h$ and integrating by parts on each element, one arrives at [3]: Find $\mathbf{A}_h^\alpha \in V_h$ such that for all $\mathbf{A}_h' \in V_h$:

$$a_h^{\text{NWIP}}(\mathbf{A}_h^\alpha, \mathbf{A}_h') + \int_\Omega \kappa^\alpha \mathbf{A}_h^\alpha \cdot \overline{\mathbf{A}_h'} = \int_\Omega \mathbf{j}^i \cdot \overline{\mathbf{A}_h'}, \tag{12.3}$$

with sesquilinear form

$$a_h^{\text{NWIP}}(\mathbf{A}_h^\alpha, \mathbf{A}_h') := \int_\Omega \mu^{-1}\mathbf{curl}\mathbf{A}_h^\alpha \cdot \mathbf{curl}\overline{\mathbf{A}_h'} - \sum_{F \in \mathscr{F}_h^i} \int_F \left\{\!\!\left\{ \mu^{-1}\mathbf{curl}\mathbf{A}_h^\alpha \right\}\!\!\right\}_w \cdot \overline{[\![\mathbf{A}_h']\!]}_T$$

$$+ \sum_{F \in \mathscr{F}_h^i} \int_F \overline{\left\{\!\!\left\{ \mu^{-1}\mathbf{curl}\mathbf{A}_h' \right\}\!\!\right\}_w} \cdot [\![\mathbf{A}_h^\alpha]\!]_T + \sum_{F \in \mathscr{F}_h^i} \frac{\eta\gamma_{\mu,F}}{h_F} \int_F [\![\mathbf{A}_h^\alpha]\!]_T \cdot \overline{[\![\mathbf{A}_h']\!]}_T.$$

Here h_F is the diameter of face F and $\eta > 0$ is the penalty parameter. The weights for an inner face $F = \partial T_1 \cap \partial T_2$ are chosen as

$$\gamma_{\mu,F} := \frac{2}{\mu_1 + \mu_2}, \qquad w_1 := \frac{\mu_1}{\mu_1 + \mu_2}, \qquad w_2 := \frac{\mu_2}{\mu_1 + \mu_2}.$$

We have the following best approximation result, cf. [3, Theorem 3.3.13]:

Theorem 12.1 *Let $\mathbf{A}^\alpha \in V^*(P_\Omega)$ be the solution of the perturbed problem (12.2) and let $\mathbf{A}_h^\alpha \in V_h$ solve the NWIP formulation (12.3). Then there exist constants $C > 0$, $C_\eta > 0$, both independent of h, μ, κ such that for $\eta > C_\eta$*

$$\left\| \mathbf{A}^\alpha - \mathbf{A}_h^\alpha \right\|_{IP} < C \inf_{\mathbf{v}_h \in V_h} \|\mathbf{A} - \mathbf{v}_h\|_{IP,*}, \tag{12.4}$$

and the discrete problem (12.3) is well-posed. The constants C_η, C depend on the choice of the subspace $V_h \subset V_h^*$ and C_η depends on C.

The associated (semi-) norms are defined as:

$$\|\mathbf{A}\|_{\text{IP}}^2 := \left\| \mu^{-1/2}\mathbf{curl}\mathbf{A} \right\|_{L^2(\Omega)^3}^2 + \left\| \sqrt{|\kappa^\alpha|}\mathbf{A} \right\|_{L^2(\Omega)^3}^2 + \sum_{F \in \mathscr{F}_h^i} \frac{\gamma_{\mu,F}}{h_F} \left\| [\![\mathbf{A}]\!]_T \right\|_{L^2(F)^3}^2,$$

$$\|\mathbf{A}\|_{\text{IP},*}^2 := \|\mathbf{A}\|_{\text{IP}}^2 + \sum_{T \in \mathscr{T}_h} h_T \left\| \mu^{-1/2}\mathbf{curl}\mathbf{A} \right\|_{L^2(T)^3}^2.$$

12.3   Enriched Approximation Space

Trefftz Functions Let **n** be a unit vector and consider problem (12.1) on the whole space $\mathbb{R}^3$ such that σ is zero in the half-space $\Omega_0 = \{x \in \mathbb{R}^3|\ x \cdot n > 0\}$ and equal to a positive constant σ in the other half-space $\Omega_\sigma = \mathbb{R}^3 \setminus \overline{\Omega_0}$. Furthermore, assume that there is an external excitation by a magnetic field $\mathbf{H}_0$ which is constant along the surface $F := \{x \in \mathbb{R}^3|\ x \cdot n = 0\}$ and that $\mu \equiv const$, $\mathbf{j}^i = 0$ in Ω_σ. Simple manipulations (cf. [5]) show that inside the conductor Ω_σ ($x \cdot n < 0$) we can write the solution **A** of (12.1) explicitly as

$$\mathbf{A}(x) = \mathbf{A}_{F,\tau}(x) := |\mathbf{H}_0|\, \delta/(1+i)\ \boldsymbol{\tau}\ \exp\left((1+i)(x - x_0) \cdot n/\delta\right), \qquad (12.5)$$

where $x_0 \in F$, $\boldsymbol{\tau} \in \mathbb{R}^3$ is a vector tangential to F, and $\delta = \sqrt{\frac{2}{\mu \sigma \omega}}$ is the skin-depth.

Modulated Trefftz Functions Let $\mathbb{P}_k(T)$ denote the space of polynomials of total degree $\le k$ on mesh element $T \in \mathcal{T}_h$. For each element $T \in \mathcal{T}_h$, $T \subset \Omega_\sigma$ we define the space

$$\mathscr{A}_k(T) := \left\{ p \mathbf{A}_{F,\tau}|\ p \in \mathbb{P}_k(T),\ F \in \mathcal{F}_h^i,\ F \subset \partial T \cap \partial \Omega_0,\ \boldsymbol{\tau}\ \text{tangential of } F \right\}.$$

Note that the dimension of the space $\mathscr{A}_k(T)$ is $2n \dim(\mathbb{P}_k(T))$, where n is the number of faces of T that are at the conductor surface, since for every flat surface there are only two linearly independent tangentials $\boldsymbol{\tau}$. We define $\mathcal{T}_h^{\mathscr{A}} := \{T \in \mathcal{T}_h|\ \dim(\mathscr{A}_1(T)) > 0, \sigma(T) > 0\}$ to be the set of elements with at least one adjacent boundary layer and we let $\Omega_{\mathscr{A}} \subset \Omega$ be the union of all elements in $\mathcal{T}_h^{\mathscr{A}}$. We then define the *broken, modulated Trefftz approximation space* by

$$\mathscr{A}_k(\mathcal{T}_h) := \left\{ \mathbf{A} \in L^2(\Omega_{\mathscr{A}})^3|\ \mathbf{A}|_T \in \mathscr{A}_k(T)\ \forall T \in \mathcal{T}_h^{\mathscr{A}} \right\}.$$

Broken Edge Element Space Our idea is to use a conforming edge element space wherever possible and to "break" this space only around elements containing the modulated Trefftz functions:

$$R_{k,\mathscr{A}}(\mathcal{T}_h) := \Big\{ \mathbf{A} \in L^2(\Omega)^3|\ \mathbf{A}|_T \in R_k(T)\ \forall T \in \mathcal{T}_h,\ n \times \mathbf{A} = 0 \text{ on } \partial\Omega,$$

$$\text{and}\ \ \mathbf{A}|_{\Omega \setminus \overline{\Omega_{\mathscr{A}}}} \in \mathbf{H}(\mathbf{curl};\ \Omega \setminus \overline{\Omega_{\mathscr{A}}}) \Big\}.$$

Here $R_k(T)$ is the space of k-th order edge elements of the first kind on mesh element $T \in \mathcal{T}_h$, cf. [2, 7] and $\mathbf{H}(\mathbf{curl};\ \Omega) := \{\mathbf{A} \in L^2(\Omega)^3|\ \mathbf{curl}\mathbf{A} \in L^2(\Omega)^3\}$. We define the enriched approximation space V_h on mesh $\mathcal{T}_h$ as

$$V_h := R_{k,\mathscr{A}}(\mathcal{T}_h) \oplus \mathscr{A}_k(\mathcal{T}_h).$$

Note that this space is tangentially continuous across a face $F \in \mathscr{F}_h^i$ if and only if both the adjacent elements do *not* belong to $\mathscr{T}_h^{\mathscr{A}}$. I.e. the DG-terms on these faces drop out of the NWIP formulation (12.3) and the method resembles "locally" the standard finite element method. Moreover we note that V_h is a superset of the space of conforming edge elements, $R_{k,h} := \{ \mathbf{A} \in \mathbf{H}(\mathbf{curl}; \Omega) |\ \mathbf{A}|_T \in R_k(T)\ \forall T \in \mathscr{T}_h \}$. In light of the best approximation result (12.4) we can thus expect that the space V_h has equal or better approximation properties than the space $R_{k,h}$.

12.4 Numerical Example

We pose problem (12.2) on a cylindrical shaped domain Ω with two conductors Ω_σ: The "plate" Ω_{plate} (green) is the cuboid $(-0.7, -0.5) \times (-1, 1)^2$ whereas the "bar" (gray) has dimensions $(0.5, 1.5) \times (-2.5, 2.5) \times (-0.5, 0.5)$. We mesh Ω with the coarse, hybrid mesh $\mathscr{T}_h$ shown in Fig. 12.1 that has only one layer of elements across the plate. This reflects the constraints encountered with more complex geometries where it is prohibitively expensive to resolve the boundary layers with a fine mesh.

The system is excited by a homogeneous generator current, $\mathbf{j}^i = (0, 2000, 0)$ in Ω_{bar}, which induces an electric current in the plate. We will vary σ_{plate} to simulate boundary layers of arbitrary thickness in the plate and keep all other (material) parameters constant: $\mu \equiv 4\pi \cdot 10^{-7}$ globally, $\sigma_{\text{bar}} = 10^4$, $\omega = 50$, and $\alpha = 10^{-6}$.

Figure 12.2 shows a first, qualitative comparison of the current distribution in a cross section of the plate. Comparing the reference solution[1] with the solution of

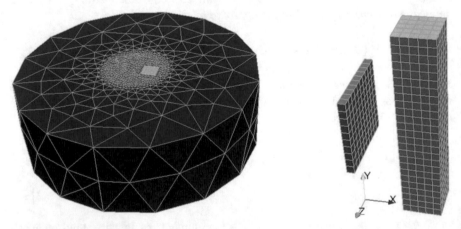

Fig. 12.1 Coarse, hybrid mesh of domain Ω with airbox (left) and without airbox (right), $h = 0.2$

---

[1]The reference solution was obtained on an refined mesh, which is adapted to the local features of the solution, using second order edge elements.

Fig. 12.2 Current distribution $|\mathbf{j}| = |\omega\sigma\mathbf{A}^\alpha|$ in plate plotted over cross-section $y = 0$ for $\sigma_{\text{plate}} = 5 \cdot 10^7$, $\delta_{\text{plate}}/h = 0.063$

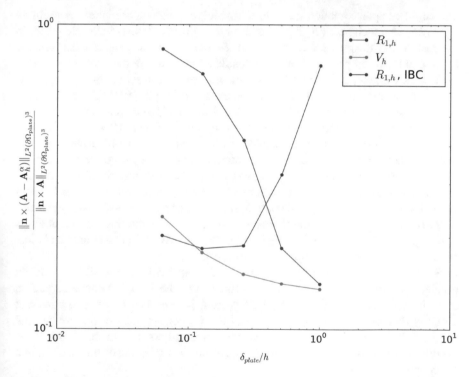

Fig. 12.3 Local surface error vs. skin-depth δ for the mesh shown in Fig. 12.1

the standard, first order FEM, we see that the top and bottom boundary layers are not resolved at all and the behavior in the edges is completely wrong. The proposed (modulated) Trefftz method with $k = 1$ can resolve the bottom and top boundary layer much better but the error is still considerable in the edges.

Figure 12.3 shows the *relative* local surface error $\dfrac{\|\mathbf{n}\times(\mathbf{A}-\mathbf{A}_h^\alpha)\|_{L^2(\partial\Omega_{\text{plate}})^3}}{\|\mathbf{n}\times\mathbf{A}\|_{L^2(\partial\Omega_{\text{plate}})^3}}$ for different values of σ_{plate} (and hence δ_{plate}). We observe that the error of the enriched method is always equal or better than simple first-order edge functions $R_{1,h}$. In particular for $\delta \ll h$ the modulated Trefftz functions clearly outperform the classical edge elements, cf. Fig. 12.2. For reference we also show the error for a standard, first-order FEM formulation where the plate has been replaced by IBC [8]. We see that the IBC approximation becomes valid as $\delta_{plate} \to 0$ and does in fact reach the precision of the enriched method for small δ. The former is expected since for flat surfaces the IBC solution tends to $\mathbf{A}$ with rate $O(\delta^2)$ [6].

12.5 Concluding Remarks

The enriched approximation space V_h can resolve the boundary layers of problem (12.1) *locally* much better than pure, standard first order Nédélec/edge elements. In contrast to IBC, the presented method also resolves the electromagnetic fields inside of the conductor. In particular, it is applicable to cases where the excitation current $\mathbf{j}^i$ generates boundary layers.[2] We remark that the construction of the functions $\mathbf{A}_{F,\tau}$ is based on the same principle that is used to derive the IBC [8]. In particular, both methods perform very well along flat surfaces but lead to considerable error in edges/corners of the geometry where the assumptions of Sect. 12.3 become invalid and the solution shows singular behavior. A more extensive numerical study unveils that the smaller δ, the more the approximation error $\mathbf{A}^\alpha - \mathbf{A}_h^\alpha$ is concentrated in the edges/corners of the plate. I.e. the approximation error is dominated by the error at corners/edges and choosing a higher order of approximation, $k > 1$ in V_h, will generally not improve the approximation. Instead one has to resolve the singularities either by refining the mesh towards edges/corners or by including the singularities in the approximation space V_h. The latter is particularly attractive since this is just another "enrichment" of the approximation space V_h.

However, finding explicit expressions for the singularities of the 3D eddy current problem at corner points is extremely difficult. For the 2D eddy current equation explicit expressions for these singularities exist [4] and can be used to construct a highly efficient method that shows exponential convergence in the polynomial degree k *independent of* δ, that is the method is *robust* in δ in the sense of [9, Definition 3.54]. We will present the details of our investigation of this method in a future work.

Acknowledgements This work has be co-funded by the Swiss Commission for Technology and Innovation (CTI).

---

[2]This is confirmed by numerical experiments not shown in this work.

References

1. Bachinger, F., Langer, U., Schöberl, J.: Numerical analysis of nonlinear multiharmonic eddy current problems. Technical report, Johannes Kepler University Linz, 2004. SFB-Report No. 2004-01
2. Bergot, M., Duruflé, M.: High-order optimal edge elements for pyramids, prisms and hexahedra. J. Comput. Phys. **232**(1), 189–213 (2013)
3. Casagrande, R.: Discontinuous finite element methods for eddy current simulation. Ph.D. thesis, ETH Zürich (2017)
4. Dauge, M., Dular, P., Krähenbühl, L., Péron, V., Perrussel, R., Poignard, C.: Corner asymptotics of the magnetic potential in the eddy-current model. Math. Methods Appl. Sci. **37**(13), 1924–1955 (2014)
5. Jackson, J.D.: Classical Electrodynamics. Wiley, London (1999)
6. Mitzner, K.: An integral equation approach to scattering from a body of finite conductivity. Radio Sci. **2**(12), 1459–1470 (1967)
7. Monk, P.: Finite Element Methods for Maxwell's Equations. Oxford University Press, Oxford (2003)
8. Ostrowski, J.: Boundary element methods for inductive hardening. Ph.D. thesis, Universität Tübingen (2002)
9. Schwab, C.: p-and hp-Finite Element Methods: Theory and Applications in Solid and Fluid Mechanics. Oxford University Press, Oxford (1998)
10. Xenophontos, C.: The hp finite element method for singularly perturbed problems in nonsmooth domains. Numer. Methods Partial Differ. Equ. **15**(1), 63–90 (1999)

Chapter 13
Survey on Semi-explicit Time Integration of Eddy Current Problems

Jennifer Dutiné, Markus Clemens, and Sebastian Schöps

Abstract The spatial discretization of the magnetic vector potential formulation of magnetoquasistatic field problems results in an infinitely stiff differential-algebraic equation system. It is transformed into a finitely stiff ordinary differential equation system by applying a generalized Schur complement. Applying the explicit Euler time integration scheme to this system results in a small maximum stable time step size. Fast computations are required in every time step to yield an acceptable overall simulation time. Several acceleration methods are presented.

13.1 Introduction

Spatially discretizing the magnetic vector potential formulation of eddy current problems, e.g by the Finite Element Method (FEM), yields a differential-algebraic equation system (DAE) [1]. Commonly, only unconditionally stable implicit time integration methods as e.g. the implicit Euler method or the singly diagonal implicit Runge-Kutta schemes can be used for time integration of the infinitely stiff DAE system [2]. In every time step at least one large nonlinear algebraic equation system needs to be solved due to the nonlinear BH-characteristic in ferromagnetic materials. The Newton-Raphson method is frequently used for linearization and requires at least one iteration per time step. Here, the Jacobian and the resulting stiffness matrix are updated in each iteration and the resulting linear algebraic equation system needs to be solved efficiently.

Applying explicit time integration schemes avoids the necessity of linearization, because nonlinearities only appear in right-hand side expressions. A first approach to use an explicit time integration method for eddy current problems has been

J. Dutiné (✉) · M. Clemens
University of Wuppertal, Wuppertal, Germany
e-mail: dutine@uni-wuppertal.de; clemens@uni-wuppertal.de

S. Schöps
Technische Universität Darmstadt, Graduate School CE, Darmstadt, Germany
e-mail: schoeps@temf.tu-darmstadt.de

© Springer International Publishing AG, part of Springer Nature 2018
U. Langer et al. (eds.), *Scientific Computing in Electrical Engineering*,
Mathematics in Industry 28, https://doi.org/10.1007/978-3-319-75538-0_13

proposed in [3], where in the conducting regions of the problems the Finite Difference Time Domain (FDTD) method is used. In the nonconducting regions, i.e., in the air, the corresponding parts of the solution are computed using the boundary element method (BEM) [3]. In a second approach presented in [4], the Discontinuous Galerkin FEM and an explicit time integration method are used for computations in the conducting regions. Continuous FEM ansatz functions and an implicit time integration scheme are applied to the nonconducting regions of the problem [4]. Both approaches in [3] and [4] are based on a separate treatment of conducting and nonconducting regions. A different approach presented in [1] and [5] proposes a Schur complement reformulation of the eddy current problem. In [6] the use of a generalized Schur complement is proposed. Here, a pseudo-inverse of the singular curl-curl matrix in nonconducting regions is evaluated using the preconditioned conjugate gradient (PCG) method. This evaluation forms a multiple-right hand side problem and suitable start vectors for the PCG method are computed using the cascaded Subspace Projection Extrapolation (CSPE) method, which is a modification of the Subspace Projection Extrapolation (SPE) method [6, 7]. Alternatively, the Proper Orthogonal Decomposition (POD) method can be used for computing improved start vectors [8]. Computations can be accelerated further by using a selective update strategy for updating the reluctivity matrix in conducting regions [9]. This paper presents a survey on the methods presented in [6, 8, 9].

13.2   Mathematical Formulation

The partial differential equation

$$\kappa \frac{\partial \mathbf{A}(t)}{\partial t} + \nabla \times (\nu \, (\mathbf{A}(t)) \, \nabla \times \mathbf{A}(t)) = \mathbf{J}_s(t), \tag{13.1}$$

describes magnetoquasistatic field problems using the time-dependent magnetic vector potential $\mathbf{A}(t)$, where κ is the electrical conductivity, ν is the eventually ferromagnetic, i.e., nonlinearly field dependent, reluctivity and $\mathbf{J}_s(t)$ is the transient source current density.

Discretizing (13.1) in space, e.g. by FEM using edge elements [10, 11], yields a differential-algebraic equation system (DAE) described by

$$\mathbf{M} \frac{d}{dt} \mathbf{a} + \mathbf{K}(\mathbf{a})\mathbf{a} = \mathbf{j}_s, \tag{13.2}$$

where $\mathbf{M}$ is the mass-matrix, $\mathbf{a}$ is the time dependent vector of the magnetic vector potential, $\mathbf{K}$ is the stiffness-matrix and $\mathbf{j}_s$ is the vector of the transient source currents. The degrees of freedom (DoFs) are separated into two vectors $\mathbf{a}_c$ and $\mathbf{a}_n$ for conducting and nonconducting material, respectively and (13.2) is re-ordered into

$$\begin{pmatrix} \mathbf{M}_{cc} & 0 \\ 0 & 0 \end{pmatrix} \frac{d}{dt} \begin{pmatrix} \mathbf{a}_c \\ \mathbf{a}_n \end{pmatrix} + \begin{pmatrix} \mathbf{K}_{cc}\,(\mathbf{a}_c) & \mathbf{K}_{cn} \\ \mathbf{K}_{cn}^{\mathsf{T}} & \mathbf{K}_{nn} \end{pmatrix} \begin{pmatrix} \mathbf{a}_c \\ \mathbf{a}_n \end{pmatrix} = \begin{pmatrix} 0 \\ \mathbf{j}_{s,n} \end{pmatrix}, \tag{13.3}$$

where $\mathbf{M}_{cc}$ is the conductivity matrix in conducting regions, $\mathbf{K}_{cc}\,(\mathbf{a}_c)$ is the nonlinear part of the reluctivity related stiffness matrix in conducting regions, $\mathbf{K}_{nn}$ is the part of the curl-curl operator in air, which is singular, and $\mathbf{j}_{s,n}$ is the source current vector corresponding to excitations in nonconducting regions. $\mathbf{M}_{cc}$ is positive definite if using a conventional Galerkin scheme with (possibly high-order) edge elements as test and ansatz functions [10, 11].

The generalized Schur complement expression

$$\mathbf{K}_S := \mathbf{K}_{cn}\mathbf{K}_{nn}^{+}\mathbf{K}_{cn}^{\mathsf{T}}, \tag{13.4}$$

where $\mathbf{K}_{nn}^{+}$ is the matrix representation of a pseudo-inverse of $\mathbf{K}_{nn}$, is applied to (13.3) and transforms the DAE into an ordinary differential equation (ODE) system

$$\mathbf{M}_{cc}\frac{d}{dt}\mathbf{a}_c + (\mathbf{K}_{cc}\,(\mathbf{a}_c) - \mathbf{K}_S)\,\mathbf{a}_c = -\mathbf{K}_{cn}\mathbf{K}_{nn}^{+}\mathbf{j}_{s,n}, \tag{13.5}$$

$$\mathbf{a}_n = \mathbf{K}_{nn}^{+}\mathbf{j}_{s,n} - \mathbf{K}_{nn}^{+}\mathbf{K}_{cn}^{\mathsf{T}}\mathbf{a}_c, \tag{13.6}$$

for the vector $\mathbf{a}_c$, i.e., the degrees of freedom only situated in conductive material [1, 5, 6]. The preconditioned conjugate gradient (PCG) method is used for evaluating a pseudo-inverse of $\mathbf{K}_{nn}$ [6]. Alternatively, the singular matrix $\mathbf{K}_{nn}$ can be regularized using a grad-div regularization by which $\mathbf{K}_{nn}$ is transformed into the discrete Laplacian operator in free space [5]. Due to finite stiffness, (13.5) can be integrated in time using explicit time integration schemes as e.g. the explicit Euler method. Here, in the m-th time step the expressions

$$\mathbf{a}_c^m := \mathbf{a}_c^{m-1} + \Delta t \mathbf{M}_{cc}^{-1}\left[-\mathbf{K}_{cn}\mathbf{K}_{nn}^{+}\mathbf{j}_{s,n}^m - \left(\mathbf{K}_{cc}(\mathbf{a}_c^{m-1}) - \mathbf{K}_S\right)\mathbf{a}_c^{m-1}\right], \tag{13.7}$$

$$\mathbf{a}_n^m := \mathbf{K}_{nn}^{+}\mathbf{j}_{s,n}^m - \mathbf{K}_{nn}^{+}\mathbf{K}_{cn}^{\mathsf{T}}\mathbf{a}_c^m, \tag{13.8}$$

are evaluated for the degrees of freedom in the conductor domain and in the nonconductive domains consecutively, where Δt is the time step size.

Evaluating a pseudo-inverse of $\mathbf{K}_{nn}$ and the inverse of $\mathbf{M}_{cc}$ in (13.7) and (13.8) repeatedly using the PCG method forms multiple right-hand side (mrhs) problems since the matrices involved remain constant. The subspace extrapolation (SPE) method can be used for computing improved start vectors for the PCG method [6, 7]. Solution vectors from n previous time steps are orthonormalized using the modified Gram-Schmidt method and form the linearly independent column vectors of the operator $\mathbf{V}$. The projected system

$$\mathbf{V}^{\mathsf{T}}\mathbf{K}_{nn}\mathbf{V}\mathbf{z} = \mathbf{V}^{\mathsf{T}}\mathbf{r}, \tag{13.9}$$

where $\mathbf{r}$ represents the new right-hand side for the full system, is solved for $\mathbf{z} \in \mathbb{R}^n$ using a direct method. The linear combination of the column vectors in $\mathbf{V}$ weighted with the coefficients in $\mathbf{z}$ yields the improved start vector $\mathbf{x}_{0,\text{CSPE}}$:

$$\mathbf{x}_{0,\text{CSPE}} := \mathbf{Vz}. \tag{13.10}$$

Only the last column vector in the operator $\mathbf{V}$ changes in every time step. Therefore, when computing $\mathbf{K}_{nn}\mathbf{V}$ in (13.9), all other matrix-column-vector products evaluated can be reused from previous time steps. This modification of the SPE start vector generation method is referred to as "cascaded SPE" (CSPE).

Alternatively, the proper orthogonal decomposition (POD) method can be used for computing improved start vectors for the PCG method [8, 12]. A snapshot matrix $\mathbf{X}$ is assembled using solutions from previous time steps as column vectors. This matrix is decomposed by the singular value decomposition (SVD) [13] into:

$$\mathbf{X} = \mathbf{U}\boldsymbol{\Sigma}\mathbf{V}^\top, \tag{13.11}$$

where $\boldsymbol{\Sigma}$ is a diagonal matrix of the singular values and $\mathbf{U}$ and $\mathbf{V}$ are orthogonal matrices. The first k column vectors of $\mathbf{U}$ corresponding to the k largest singular values $\sigma_1, \ldots, \sigma_k$, for which holds

$$\sigma_i \geq \sigma_j, \text{ for } i < j, \tag{13.12}$$

$$\frac{\sigma_1}{\sigma_k} \geq tol_{\text{POD}}, \tag{13.13}$$

become the column vectors of the reduced matrix $\mathbf{U}_r$ with a threshold value tol_{POD}. A threshold value commonly used in practical computations is $tol_{\text{POD}} := 10^4$. The improved start vector $\mathbf{x}_{0,\text{POD}}$ for the PCG method is computed by

$$\mathbf{x}_{0,\text{POD}} := \mathbf{U}_r \left[\mathbf{U}_r^\top \mathbf{K}_{nn} \mathbf{U}_r \right]^{-1} \mathbf{U}_r^\top \mathbf{K}_{cn}^\top \mathbf{a}_c. \tag{13.14}$$

The explicit Euler method is only stable for time step sizes Δt smaller than a Courant-Friedrich-Levy-type time step size Δt_{CFL} given by [1]:

$$\Delta t_{\text{CFL}} \leq \frac{2}{\lambda_{\max}\left(\mathbf{M}_{cc}^{-1} \left(\mathbf{K}_{cc}\left(\mathbf{a}_c \right) - \mathbf{K}_S \right) \right)}, \tag{13.15}$$

where the maximum eigenvalue $\lambda_{\max}$ is proportional to

$$\lambda_{\max}\left(\mathbf{M}_{cc}^{-1} \left(\mathbf{K}_{cc}\left(\mathbf{a}_c \right) - \mathbf{K}_S \right) \right) \propto \frac{1}{h^2 \kappa \mu}, \tag{13.16}$$

assuming non-singularity of $\mathbf{M}_{cc}$. Here, h is the smallest edge length in the mesh, κ is the electrical conductivity, and μ is the permeability. Numerical tests have shown that $1/(h^2 \kappa \mu)$ unfortunately does not give a sharp estimate of $\lambda_{\max}$, such that the largest eigenvalue has to be computed numerically.

Fine spatial discretization can result in small stable time step sizes, due to (13.15), that can be in the micro- to nano second range. Considering the dynamics of the usual excitation currents in magnetoquasistatic problems, this corresponds to a strong over-sampling. It is assumed that the excitation current does not change significantly between succeeding time steps. Therefore it is expected that the vector $\mathbf{a}_c$ in (13.7), (13.8) also only changes marginally between succeeding time steps. The matrix $\mathbf{K}_{cc}(\mathbf{a}_c)$ is thus only updated if the change between the vector $\mathbf{a}_c^m$ at the time step m and the vector $\mathbf{a}_c^l$ from the time step $l < m$ at which the matrix $\mathbf{K}_{cc}(\mathbf{a}_c^l)$ has last been updated is larger than a chosen tolerance tol, as described by [9]:

$$\frac{\|\mathbf{a}_c^m - \mathbf{a}_c^l\|}{\|\mathbf{a}_c^l\|} > tol, \tag{13.17}$$

where $\|\cdot\|$ denotes the l2-norm. However, depending on the gauging used, a different norm might be more appropriate, e.g. using the magnetic energy norm.

13.3 Numerical Validation

The ferromagnetic TEAM 10 benchmark problem is spatially discretized using first order edge element FEM ansatz functions [14, 15]. The model geometry is shown in Fig. 13.1. The excitation current is described by a $(1 - \exp(-t/\tau))$ function. A time interval of 120 ms duration is calculated. The accuracy of the employed simulation code is proven using an implicit time integration method and a fine mesh discretization of about 700,000 DoFs. The resulting average magnetic flux density is compared with the measurement results published in [14] in Fig. 13.2. As this simulation takes a simulation time of 5.38 days on a workstation with an Intel Xeon

Fig. 13.1 TEAM 10 model geometry and position S1. Steel plates are colored in blue and red, the coil in green. There is a 0.5 mm wide air gap between the blue and red steel plates

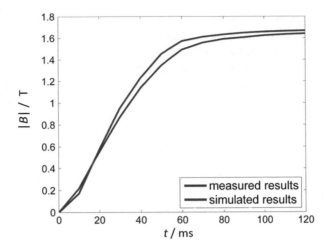

Fig. 13.2 Comparison of simulation results using a mesh of 700,000 DoFs and the measured results published in [14] at position S1

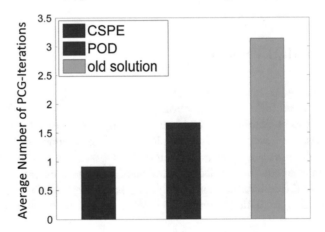

Fig. 13.3 Averagely required number of PCG iterations for evaluating the pseudo-inverse of $\mathbf{K}_{nn}$ using either CSPE, POD, or the solution from the previous time step as start vector for the PCG method

E5 processor, a coarser mesh is applied for the simulations using the explicit Euler method for time integration. The applied coarse spatial discretization yields 29,532 DoFs and results in a maximum stable time step size $\Delta t_{CFL} = 1.2\,\mu s$, such that 100,000 explicit Euler time steps are required for this problem.

Computing improved start vectors for the PCG method using either CSPE or POD reduces the average number of required PCG iterations compared to using the solution from the previous time step as start vector. An algebraic multigrid method is used as preconditioner. The results for the evaluation of the pseudo-inverse of $\mathbf{K}_{nn}$ using a PCG tolerance of 10^{-6} are shown in Fig. 13.3. Using the selective update strategy for updating the matrix $\mathbf{K}_{cc}$ $(\mathbf{a}_c)$ does not significantly decrease accuracy, as is shown in Fig. 13.4. The number of required updates and the simulation time are significantly reduced, as is depicted in Figs. 13.5 and 13.6. If $\mathbf{K}_{cc}$ $(\mathbf{a}_c)$ is updated in every time step 100,000 updates are performed during the entire simulation. A workstation with an Intel Xeon E5 processor and an NVIDIA TESLA K80 GPU are used for these simulations. The matrix $\mathbf{M}_{cc}$ is inverted directly using GPU acceleration. This is only possible, as the matrix $\mathbf{M}_{cc}$ is only of dimension 5955×5955 in this test problem. For more refined discretizations the PCG method should be used for inverting the matrix $\mathbf{M}_{cc}$.

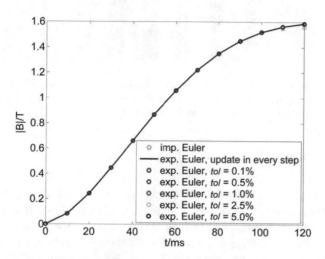

Fig. 13.4 Average magnetic flux density at position S1

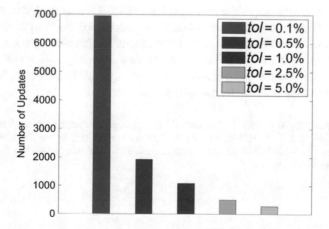

Fig. 13.5 Number of updates of $\mathbf{K}_{cc}$ $(\mathbf{a}_c)$ for different tolerances *tol* in (13.17)

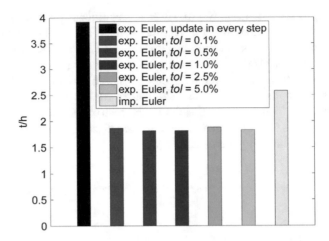

Fig. 13.6 Comparison of simulation times for a simulation using implicit Euler method, the explicit Euler method with updates of $\mathbf{K}_{cc}\,(\mathbf{a}_c)$ in every time step and the explicit Euler method using the selective update strategy and different tolerances *tol*

13.4   Conclusion

The application of a generalized Schur complement to the spatially discretized magnetic vector potential formulation of magnetoquasistatic field problems transformed a DAE of infinite stiffness into a finitely stiff system of ODEs. This ODE system is integrated with the explicit Euler method. For the evaluation of a pseudo-inverse the PCG method was adopted. Improved start vectors were computed with the CSPE and the POD method, reducing the number of required PCG iterations in simulations of the ferromagnetic TEAM 10 benchmark problem. A selective update strategy for the reluctivity matrix taking into account the specific problem dynamics reduced the number of required updates and the simulation time. So far, the small stable time step size of the explicit Euler method results in high computational effort which can be overcome using massive GPU-parallelization to reduce the required computational time per time step significantly.

Acknowledgements This work was supported by the German Research Foundation DFG (grant numbers CL143/11-1, SCHO1562/1-1). The third author is supported by the Excellence Initiative of the German Federal and State Governments and The Graduate School of Computational Engineering at TU Darmstadt.

References

1. Schöps, S., Bartel, A., Clemens, M.: Higher order half-explicit time integration of eddy current problems using domain substructuring. IEEE Trans. Magn. **48**(2), 623–626 (2012)
2. Hairer, E., Wanner, G.: Solving Ordinary Differential Equations II: Stiff and Differential-Algebraic Problems, 2nd rev. edn. Springer, Berlin (1996)
3. Yioultsis, T.V., Charitou, K.S., Antonopoulos, C.S., Tsiboukis, T.D.: A finite difference time domain scheme for transient eddy current problems. IEEE Trans. Magn. **37**(5), 3145–3149 (2001)
4. Außerhofer, S., Bíro, O., Preis, K.: Discontinuous Galerkin finite elements in time domain eddy-current problems. IEEE Trans. Magn. **45**(3), 1300–1303 (2009)
5. Clemens, M., Schöps, S., De Gersem, H., Bartel, A.: Decomposition and regularization of nonlinear anisotropic curl-curl DAEs. COMPEL **30**(6), 1701–1714 (2011)
6. Dutiné, J., Clemens, M., Schöps, S., Wimmer, G.: Explicit time integration of transient eddy current problems. Int J Numer Model. e2227 (2017) https://doi.org/10.1002/jnm.2227
7. Clemens, M., Wilke, M., Schuhmann, R., Weiland, T.: Subspace projection extrapolation scheme for transient field simulations. IEEE Trans. Magn. **40**(2), 934–937 (2004)
8. Dutiné, J., Clemens, M., Schöps, S.: Multiple right-hand side techniques in semi-explicit time integration methods for transient eddy-current problems. IEEE Trans. Magn. **53**(6), 1–4 (2017)
9. Dutiné, J., Clemens, M., Schöps, S.: Explicit time integration of eddy current problems using a selective matrix update strategy. Compel **36**(5), 1364–1371 (2017)
10. Monk, P.: Finite Element Methods for Maxwell's Equations. Oxford University Press, Oxford (2003)
11. Jin, J.-M.: The Finite Element Method in Electromagnetics, 3rd edn. Wiley-IEEE Press, Hoboken (2014)
12. Chatterjee, A.: An introduction to the proper orthogonal decomposition. Curr. Sci. **78**(7), 808–817 (2000)
13. Trefethen, L.N., Bau, D.: Numerical Linear Algebra. Society for Industrial and Applied Mathematics, Philadelphia (1997)
14. Nakata, T., Fujiwara, K.: Results for benchmark problem 10 (steel plates around a coil). COMPEL **9**(3), 181–192 (1990)
15. Kameari, A.: Calculation of transient 3D eddy current using edge-elements. IEEE Trans. Magn. **26**(2), 466–469 (1990)

Chapter 14
A Local Mesh Modification Strategy for Interface Problems with Application to Shape and Topology Optimization

Peter Gangl and Ulrich Langer

Abstract We present and analyze a new finite element method for solving interface problems on a triangular grid. The method locally modifies a given triangulation such that the interfaces are accurately resolved and a maximum angle condition holds. Therefore, optimal order of convergence can be shown. Moreover, it can be shown that an appropriate choice of the basis functions yields an optimal condition number of the stiffness matrix. The method is applied to an optimal design problem for an electric motor where the interface between different materials evolves in the course of the optimization procedure.

14.1 Motivation

Our research is motivated by the design optimization of an electric motor by means of topology and shape optimization. We are interested in finding the optimal distribution of two materials (usually ferromagnetic material and air) within a fixed design subdomain of an electric motor, see, e.g. [1]. We employ a two-dimensional model for the electric motor, which is widely used for this kind of applications. In the optimization procedure, one usually starts with an initial guess and then uses shape sensitivities or topological sensitivities to gradually improve the initial design. In the course of this optimization procedure, the interface between the two subdomains evolves. For computing the sensitivities that steer the optimization process, it is necessary to solve the state equation and the adjoint equation in each

P. Gangl (✉)
Institut für Numerische Mathematik, TU Graz, Graz, Austria

Linz Center of Mechatronics GmbH (LCM), Linz, Austria
e-mail: gangl@math.tugraz.at

U. Langer
Institute of Computational Mathematics (NuMa), JKU Linz, Linz, Austria
e-mail: ulanger@numa.uni-linz.ac.at

© Springer International Publishing AG, part of Springer Nature 2018
U. Langer et al. (eds.), *Scientific Computing in Electrical Engineering*,
Mathematics in Industry 28, https://doi.org/10.1007/978-3-319-75538-0_14

optimization iteration, which is usually done by the finite element method. Besides remeshing in every iteration, which is very costly, and advecting the whole mesh in every step of the optimization procedure, which may cause self-intersection of the mesh, there exist several other methods in the literature which can deal with these kinds of interface problems. We mention the XFEM, which uses local enrichment of the finite element basis, and the unfitted Nitsche method. In [2], shape optimization is performed by advecting the finite element mesh in combination with adaptive mesh refinement, which optimizes the required computational effort.

In [3], the authors introduce a locally modified parametric finite element method based on a quadrilateral mesh with a patch structure. We present an adaptation of this method to the case of finite elements on triangular meshes. One advantage of this kind of method over the ones mentioned before is that this method has a fixed number of unknowns independently of the position of the interface relative to the mesh. The given mesh is modified only locally near the material interface. The method is relatively easy to implement and we can show optimal order of convergence.

14.2 A Local Mesh Modification Strategy for Interface Problems

We introduce the method for the potential equation in a bounded, polygonal computational domain $\Omega \subset \mathbb{R}^2$ consisting of two non-overlapping subdomains, $\overline{\Omega} = \overline{\Omega}_1 \cup \overline{\Omega}_2$, $\Omega_1 \cap \Omega_2 = \emptyset$, which represent two materials with different material coefficients $\kappa_1, \kappa_2 > 0$. On the material interface $\Gamma := \overline{\Omega}_1 \cap \overline{\Omega}_2$, we have to require that the solution as well as the flux are continuous. For simplicity, we assume Dirichlet boundary conditions on $\partial \Omega$. The problem reads as follows:

$$-\text{div } (\kappa_i \nabla u) = f \quad \text{in } \Omega_i, \quad i = 1, 2,$$

$$[u] = 0 \quad \text{on } \Gamma,$$

$$\left[\kappa \frac{\partial u}{\partial n} \right] = 0 \quad \text{on } \Gamma, \tag{14.1}$$

$$u = g_D \quad \text{on } \partial \Omega,$$

where we assume that the boundaries of the two subdomains as well as the right hand side f and the Dirichlet data g_D are sufficiently regular such that $u \in H_0^1(\Omega) \cap H^2(\Omega_1 \cup \Omega_2)$, that means that the restrictions of $u \in H_0^1(\Omega)$ to Ω_1 and Ω_2 belong to $H^2(\Omega_1)$ and $H^2(\Omega_2)$, respectively, see, e.g., [4]. It is well-known that, when using standard finite element methods, the interface must be resolved by the mesh in order to obtain optimal convergence rates of the approximate solution u_h to the true solution u in the L^2 and H^1 norms as the mesh parameter h tends to zero, see also [3]. The discretization error estimate is usually shown using an interpolation error estimate. A condition that is sufficient and necessary for such an interpolation error estimate is that all interior angles of triangles of the mesh are bounded away from 180° (maximum angle condition), see [5].

14.2.1   Preliminaries

Let $\mathscr{T}_h$ be a shape-regular and quasi-uniform subdivision of Ω into triangular elements, and let us denote the space of globally continuous, piecewise linear functions on $\mathscr{T}_h$ by V_h. We assume that $\mathscr{T}_h$ has been obtained by one uniform refinement of a coarser mesh $\mathscr{T}_{2h}$. By this assumption, $\mathscr{T}_h$ has a patch-hierarchy, i.e., always four elements $T_1, T_2, T_3, T_4 \in \mathscr{T}_h$ can be combined to one larger triangle $T \in \mathscr{T}_{2h}$. We will refer to this larger element as the macro element or patch. We assume further that the mesh of macro elements $\mathscr{T}_{2h}$ is such that, for each macro element T, the interface Γ either does not intersect the interior of T, or such that Γ intersects T in exactly two distinct edges or that it intersects T in one vertex and in the opposite edge. For a smooth enough interface Γ, this assumption can always be enforced by choosing a fine enough macro mesh $\mathscr{T}_{2h}$. We consider a macro element $T \in \mathscr{T}_{2h}$ to be cut by the interface if the intersection of the interior of the macro element with the interface is not the empty set.

14.2.2   Description of the Method

The method presented in this paper is a local mesh adaptation strategy, meaning that only macro elements close to the interface Γ will be modified. Given the hierarchic structure of the mesh, on every macro element we have four elements of the mesh $\mathscr{T}_h$ and six vertices, see Fig. 14.1a, b. The idea of the method is the following: For each macro element that is cut by the interface, move the points P_4, P_5 and P_6 along the corresponding edges in such a way that, on the one hand, the interface is resolved accurately, and, on the other hand, all interior angles in the four triangles are bounded away from 180°. For a macro element T that is cut by the interface, we distinguish four different configurations as follows:

In the case where the macro element is cut by the interface in two distinct edges, we denote the vertex of the macro element where these two edges meet by P_1, and the other two vertices in counter-clockwise order by P_2 and P_3. The parameters s, $t, r \in [0, 1]$ represent the positions of the points P_4, P_5, P_6 along the corresponding edges by

$$P_4(s) = P_1 + s \frac{P_2 - P_1}{|P_2 - P_1|}, \quad P_5(t) = P_2 + t \frac{P_3 - P_2}{|P_3 - P_2|}, \quad P_6(r) = P_1 + r \frac{P_3 - P_1}{|P_3 - P_1|}.$$

The parameters r and s will always be chosen in such a way that the intersection points of the interface and the edges $P_1 P_3$ and $P_1 P_2$ are the points P_6 and P_4,

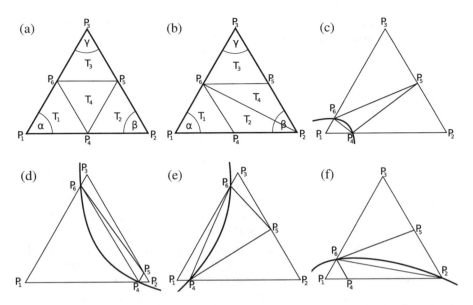

Fig. 14.1 (**a**) Patch for configurations A–C. (**b**) Patch for configuration D. (**c**) Configuration A.
(**d**) Configuration B. (**e**) Configuration C. (**f**) Configuration D

respectively. Thus, we identify the position of the interface relative to the macro
element T by the two parameters r, s. We choose the parameter t such that a
maximum angle condition is satisfied as follows:

Configuration A: $0 < r, s \leq 1/2$. Set $t = 1/2$.
Configuration B: $1/2 < r, s < 1$. Set $t = 1 - s$.
Configuration C: $0 < s \leq 1/2 < r < 1$ or $0 < r \leq 1/2 < s < 1$. Set $t = 1/2$.

The case where the macro element is cut in one vertex and the opposite edge has
to be considered separately. We denote the vertex of the macro element where it is
cut by the interface by P_2 and the other vertices, in counter-clockwise ordering, by
P_3 and P_1, see Fig. 14.1b. The location of the interface is given by the position of
the point P_6 on the edge between P_3 and P_1. In this case, we also need to rearrange
the triangles T_2 and T_4.

Configuration D:
Configuration D1: $0 < r \leq 1/2$. Set $s = r$ and $t = 1/2$.
Configuration D2: $1/2 < r < 1$. Set $s = 1/2$ and $t = r$.

With this setting, it is possible to show the required maximum angle condition
on the reference patch $\hat{T}$ defined by the outer macro vertices $\hat{P}_1 = (0, 0)^T$, $\hat{P}_2 = (1, 0)^T$, $\hat{P}_3 = (1/2, \sqrt{3}/2)^T$.

Lemma 14.1 *All angles in triangles of the reference patch $\hat{T}$ are bounded by $150°$
independent of the parameters $r, s \in [0, 1]$.*

Proof We have to ensure for each of the four subtriangles $\hat{T}_1$, $\hat{T}_2$, $\hat{T}_3$, $\hat{T}_4$ that all of their three interior angles are not larger than 150°. In Configuration A–C, the sub-triangles $\hat{T}_1$, $\hat{T}_2$ and $\hat{T}_3$ all have one angle of 60°. Obviously, the remaining two angles are bounded from above by 120°. The same holds true for the sub-triangles $\hat{T}_1$ and $\hat{T}_3$ in Configuration D.

For three points A, B, C in $\mathbb{R}^2$, define

$$\angle(A, B, C) := \cos^{-1}\left(\frac{(A - B, C - B)}{|A - B||C - B|}\right)$$

the interior angle of the triangle with vertices A, B, C at point B.

Configuration A: For $r, s \in (0, 1/2]$, we get for the angle in point P_4 that

$$\angle(P_6, P_4, P_5) < \angle(P_1, P_4, P_5) = 180° - \angle(P_5, P_4, P_2) \leq 180° - \angle(P_5, P_1, P_2).$$

Since the reference patch $\hat{T}$ is equilateral, it holds $\angle(P_5, P_1, P_2) = \alpha/2$. Analogously, we get for the angle in point P_6 that $\angle(P_5, P_6, P_4) < 180° - \alpha/2$. It is easy to see that the angle in point P_5 increases with r, s and thus is maximized for $r = s = 1/2$, which yields that $\angle(P_4, P_5, P_6) \leq \angle(P_4(1/2), P_5, P_6(1/2)) = 180 - \beta - \gamma = \alpha$. Here we used that, for $r = s = t = 1/2$, the four sub-triangles are congruent.

Configuration B: Note that, by the special choice of s, t, in this case we have that the line going through P_4 and P_5 is parallel to the edge connecting P_1 and P_3 for all values of $s \in (1/2, 1)$. Thus, we have

$$\angle(P_4, P_5, P_6) \leq \angle(P_4, P_5, P_3) = 180° - \gamma \quad \text{and}$$

$$\angle(P_4, P_5, P_6) = 180° - \gamma - \angle(P_6, P_5, P_3)$$
$$\geq 180° - \gamma - \angle(P_6(1/2), P_5(1/2), P_3) = 180° - \gamma - \beta = \alpha.$$

The angles in P_4 and in P_6 must also be bounded from above by $180° - \alpha = 120°$.

Configuration C: We consider the case where $r \in (1/2, 1)$ and $s \in (0, 1/2]$. The reverse case is treated analogously. For the angle in the fixed point $P_5 = P_5(1/2) = (P_2 + P_3)/2$, we get the estimates

$$\angle(P_4, P_5, P_6) \leq \angle(P_4, P_5, P_3) \leq \angle((P_4(1/2), P_5, P_3) = 180° - \gamma,$$

$$\angle(P_4, P_5, P_6) \geq \angle(P_4, P_5, P_6(1/2)) \geq \angle(P_1, P_5, P_6(1/2)) = \angle(P_5, P_1, P_2) = \alpha/2.$$

Thus, the angles $\angle(P_6, P_4, P_5)$ and $\angle(P_5, P_6, P_4)$ are also bounded from above by $180° - \alpha/2$.

Configuration D: We consider only Configuration D1, the corresponding result for Configuration D2 follows analogously. Due to the choice of the parameter s, the line going through P_4 and P_6 is parallel to the edge connecting P_2 and P_3 for all values of r. We need to consider triangles T_2 and T_4. In T_2, $\angle(P_6, P_4, P_2) = 180° - \beta$ and, therefore, the other two angles are bounded by β. In T_4, we have for $r \in (0, 1/2]$ that

$$\angle(P_6, P_2, P_5) \leq \beta,$$

$$\angle(P_2, P_5, P_6) \leq \angle(P_2, P_5, (P_3 + P_1)/2) = 180 - \beta,$$

$$\angle(P_5, P_6, P_2) \leq \angle(P_3, P_6, P_4) = 180 - \gamma.$$

Finally, noting that $\alpha = \beta = \gamma = 60°$ yields the statement of the lemma.

Remark 14.1 Due to the assumption that the macro mesh is shape-regular, we obtain a maximum angle condition (with a different bound) for all triangles of the mesh $\mathcal{T}_h$.

Now we are in the position to show an a priori error estimate for the finite element solution u_h. Since we have the maximum angle condition of Lemma 14.1, for smooth functions $v \in H^2(T)$, we get the interpolation error estimates

$$\|\nabla^k(v - I_h v)\|_{L^2(T)} \leq c\, h_{T,max}^{2-k} \|\nabla^2 v\|_{L^2(T)}, \quad k = 0, 1, \tag{14.2}$$

where $I_h : H^2(T) \to V_h|_{\overline{T}}$ denotes the Lagrangian interpolation operator, c is a positive generic constant, and $h_{T,max}$ is the maximum edge length of the triangle $T \in \mathcal{T}_h$, see, e.g., [6]. In the case where the interface Γ is not polygonal but smooth with C^2 parametrization, and an element of the mesh $\mathcal{T}_h$ is intersected by Γ, the solution u is not smooth across the interface and, hence, estimate (14.2) cannot be applied. However, the same estimate with $k = 1$ was shown in [7]. These interpolation error estimates allow to show the following a priori error estimate [7].

Theorem 14.1 *Let $\Omega \subset \mathbb{R}^2$ be a domain with convex polygonal boundary, split into $\Omega = \Omega_1 \cup \Gamma \cup \Omega_2$, where Γ is a smooth interface with C^2-parametrization. We assume that Γ divides Ω in such a way that the solution u belongs to $H_0^1(\Omega) \cap H^2(\Omega_1 \cup \Omega_2)$ and satisfies the stability estimate $\|u\|_{H^2(\Omega_1 \cup \Omega_2)} \leq c_s \|f\|_{L^2(\Omega)}$. Then, for the corresponding modified finite element solution $u_h \in V_h$, we have the estimates*

$$\|\nabla(u - u_h)\|_{L^2(\Omega)} \leq C\, h\, \|f\|_{L^2(\Omega)} \quad and \quad \|u - u_h\|_{L^2(\Omega)} \leq C\, h^2\, \|f\|_{L^2(\Omega)}.$$

14.3 Condition Number

The procedure of Sect. 14.2 guarantees that no angle of the modified mesh becomes too large. However, it may happen that some angles in the triangulation get arbitrarily close to zero, which usually yields a bad condition of the finite element system matrix. This problem was also addressed in [3] for the case of quadrilateral elements, and we can adapt the procedure to the triangular case.

The idea consists in a hierarchical splitting of the finite element space $V_h = V_{2h} + V_b$ into the standard piecewise linear finite element space on the macro mesh $\mathcal{T}_{2h}$ and the space of "bubble" functions in V_b which vanish on the nodes of the macro elements. Let $\{\phi_h^1, \ldots, \phi_h^{N_h}\}$ be the nodal basis of the space V_h. Any function $v_h \in V_h$ can be decomposed into the sum of a function $v_{2h} \in V_{2h} = \text{span}\{\phi_{2h}^1, \ldots, \phi_{2h}^{N_{2h}}\}$ and a function $v_b \in V_b = \text{span}\{\phi_b^1, \ldots, \phi_b^{N_b}\}$,

$$v_h = \sum_{i=1}^{N_h} v_h^i \phi_h^i = \sum_{i=1}^{N_{2h}} v_{2h}^i \phi_{2h}^i + \sum_{i=1}^{N_b} v_b^i \phi_b^i = v_{2h} + v_b \in V_{2h} + V_b.$$

In this setting, it remains to show that the basis functions ϕ_b^i of the space V_b can be scaled in such a way that the following two conditions hold:

- There exists a constant $C > 0$ independent of h, r, s such that

$$C^{-1} \leq \|\nabla \phi_h^i\| \leq C, \quad i = 1, \ldots, N_h, \tag{14.3}$$

- There exists a constant $C > 0$ independent of h, r, s, such that for all $v_b \in V_b$

$$|v_b^i| \leq C \|\nabla v_b\|_{\mathcal{N}_i}, \quad i = 1, \ldots, N_b, \text{ with } \mathcal{N}_i = \{K \in \mathcal{T}_h : x_i \in \overline{K}\}. \tag{14.4}$$

If these two assumptions are satisfied, the usual bound on the condition number of the system matrix can be shown [3].

Theorem 14.2 *Let* $\mathbf{A}$ *be the finite element system matrix of problem* (14.1) *using the hierarchical basis* $\{\phi_{2h}^1, \ldots, \phi_{2h}^{N_{2h}}, \phi_b^1, \ldots, \phi_b^{N_b}\}$ *after the mesh modification described in Sect. 14.2.2. Assume that* (14.3) *and* (14.4) *hold. Then there exists a constant* $C > 0$, *independent of the interface location* r *and* s, *such that* $\text{cond}_2(\mathbf{A}) \leq C h^{-2}$.

14.4 Numerical Results

We tested the method described in Sect. 14.2 for problem (14.1) where $\Omega = (-1, 1)^2$, the subdomain Ω_1 is a disk of radius 0.4 centered at the point $(0.1, 0.2)^T$ with material coefficient $\kappa_1 = 1$, and $\Omega_2 = \Omega \setminus \overline{\Omega}_1$ its complement with coefficient $\kappa_2 = 10$. The right hand side as well as the Dirichlet data were chosen in such a way that the exact solution is known explicitly. The optimal order of convergence stated in Theorem 14.1 can be observed in Table 14.1. Moreover, using the hierarchical basis, the behavior stated in Theorem 14.2 can be observed for the

Table 14.1 Convergence history for the interface problem (14.1) using mesh adaptation strategy

nVerts	h	$\|u - u_h\|_{L^2}$	Rate L_2	$\|\nabla(u - u_h)\|_{L^2}$	Rate H_1	angMax	$\text{cond}_2(D^{-1}\mathbf{A})$
1089	h_0	0.214909	–	9.48898	–	138.116	269.3
4225	$h_0/2$	0.0512394	**2.0684**	4.61321	**1.04048**	143.084	1630.4
16,641	$h_0/4$	0.0126861	**2.01401**	2.29116	**1.00969**	152.223	2531.0
66,049	$h_0/8$	0.00314155	**2.0137**	1.13973	**1.00739**	149.110	11,140.9
263,169	$h_0/16$	0.000784464	**2.0017**	0.568504	**1.00345**	155.643	41,497.0

The bold values numerically confirm the statement of Theorem 14.1, which is the main message of the paper

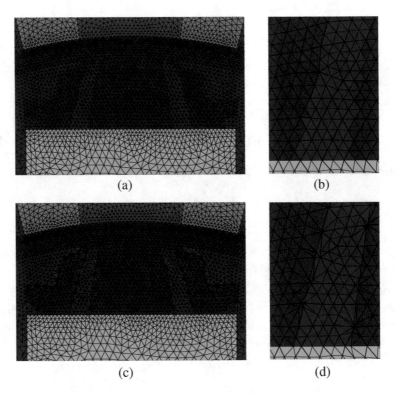

Fig. 14.2 (**a**) Final design of shape optimization without interface method, objective value $\mathscr{J}(u) \approx 0.0379$. (**b**) Zoom of (**a**). (**c**) Final design of shape optimization with interface method, objective value $\mathscr{J}(u) \approx 0.0373$. (**d**) Zoom of (**c**)

Jacobi preconditioned system matrices $D^{-1}\mathbf{A}$ with $D = \text{diag}(\mathbf{A})$. Nevertheless, conditions (14.3) and (14.4) remain to be shown.

We also included the interface method in the shape optimization of an electric motor as described in [1]. Here, the goal is to identify the shape of the ferromagnetic subdomain (brown area in Fig. 14.2) of an electric motor such that a design-dependent objective function $\mathscr{J}$, which is linked to the smoothness of the rotation of the motor, is minimized. Figure 14.2 shows the final designs obtained by the shape optimization algorithm starting out from an initial design, without (Fig. 14.2a, b) and with the proposed interface method (Fig. 14.2c, d). It can be seen that smoother and slightly better designs can be achieved by locally modifying the mesh.

Acknowledgements The authors gratefully acknowledge the Austrian Science Fund (FWF) for the financial support of their work via the Doctoral Program DK W1214 (project DK4) on Computational Mathematics. They also thank the Linz Center of Mechatronics (LCM), which is a part of the COMET K2 program of the Austrian Government, for supporting their work.

References

1. Gangl, P., Langer, U., Laurain, A., Meftahi, H., Sturm, K.: Shape optimization of an electric motor subject to nonlinear magnetostatics. SIAM J. Sci. Comput. **37**, B1002–B1025 (2015)
2. Morin, P., Nochetto, R.H., Pauletti, M.S., Verani, M.: Adaptive finite element method for shape optimization. ESAIM: COCV **18**, 1122–1149 (2012)
3. Frei, S., Richter, T.: A locally modified parametric finite element method for interface problems. SIAM J. Numer. Anal. **52**, 2315–2334 (2014)
4. Babuska, I.: The finite element method for elliptic equations with discontinuous coefficients. Computing **5**, 207–213 (1970)
5. Babuska, I., Aziz, A.K.: On the angle condition in the finite element method. SIAM J. Numer. Anal. **13**, 214–226 (1976)
6. Apel, T.: Anisotropic Finite Elements: Local Estimates and Applications. Teubner, Stuttgart (1999)
7. Frei, S.: Eulerian finite element methods for interface problems and fluid-structure-interactions. Ph.D. thesis, Universität Heidelberg (2016)

Chapter 15
Numerical Methods for Derivative-Based Global Sensitivity Analysis in High Dimensions

Qingzhe Liu and Roland Pulch

Abstract Within analysis of dynamical systems embracing uncertain impacts the output can be generally viewed as a function defined in a random domain with dependence on time or frequency. Without loss of generality, a function defined on the normalized random domain, i.e., a unit hypercube, is considered where the sensitivity analysis plays a key role in many issues, e.g. uncertainty reduction, model simplification, exploration of significant random parameters, etc. Variance-based global sensitivity indices provide adequate estimates for the influence of random variables and become one of the most powerful instruments in sensitivity analysis. Alternatively, if the function is differentiable, the derivative-based sensitivity measures have received much attention due to lower computational costs. We introduce numerical strategies for computing derivative-based sensitivity indices in the case of high-dimensional hypercubes and present numerical simulations of a test example which models the linear electric circuit of a band-stop filter.

15.1 Introduction

Sensitivity analysis of dynamical systems focuses on the study how random sources in its input impact on the output, and consequently contributes to many issues, e.g. uncertainty reduction, model order reduction, exploration of significant random parameters, calibrating model parameters, etc. in many applications of industry and finance. Especially, with the help of sensitivity analysis one can straightforward identify model parameters that cause significant uncertainty and the ones leading to tiny impacts in the response. The less important parameters can be then replaced by constants to reduce the dimensionality of the problem without a significant loss of accuracy.

Q. Liu · R. Pulch (✉)
Institut für Mathematik und Informatik, Ernst-Moritz-Arndt-Universität Greifswald, Greifswald, Germany
e-mail: pulchr@uni-greifswald.de

© Springer International Publishing AG, part of Springer Nature 2018 157
U. Langer et al. (eds.), *Scientific Computing in Electrical Engineering*, Mathematics in Industry 28, https://doi.org/10.1007/978-3-319-75538-0_15

Regarding sensitivity measures two concepts are most worth being considered: variance- and derivative-based global sensitivity indices, see e.g. [8, 10]. The variance-based global sensitivity indices have been subject to intensive research by the authors in [4, 6]. Besides, there has been a growing interest to utilize the derivative-based sensitivity measures because their relatively simple model structure causes lower computational expenses in comparison to the variance-based indices. Note that although the computation time can be counted as an advantage, investigations on differentiability of the functions in consideration are imperative prior to numerical simulations. Otherwise the simulation results should be verified by certain appropriate strategic instruments, e.g., examination of a priori estimates, comparisons with other concepts, investigations on conformance of the time and the frequency domain, etc.

With regard to numerical issues we propose two types of sampling techniques: quasi Monte Carlo (QMC) methods [7] and cubature methods [9, 12]. There is a clear distinction between these two approaches. Although QMC methods are straightforward to implement, they exhibit a relatively slow convergence rate and thus loose efficiency. In contrast, the number of samples based on cubature methods is predetermined depending on the dimensionality of the random parameter domain. Yet the cubature methods may be more efficient in comparison to QMC techniques.

The paper starts with a review of the stochastic model and derivative-based sensitivity measures. The numerical methods are specified in Sect. 15.3 where QMC methods and a cubature formula from Stroud [9] are presented. Section 15.4 illustrates the numerical performances on a linear band-stop filter as our test example while different techniques are compared.

15.2 The Stochastic Model

This section introduces linear dynamical systems, modeling of uncertainties and derivative-based sensitivity measures.

15.2.1 Linear Dynamical Systems

We consider linear single-input-single-output (SISO) dynamical systems of the form

$$\mathbf{E}(\mathbf{x})\dot{\mathbf{y}}(t, \mathbf{x}) = \mathbf{A}(\mathbf{x})\mathbf{y}(t, \mathbf{x}) + \mathbf{B}(\mathbf{x})u(t)$$
$$z(t, \mathbf{x}) = \mathbf{C}(\mathbf{x})\mathbf{y}(t, \mathbf{x}) \tag{15.1}$$

including physical parameters $\mathbf{x} = (x_1, \ldots, x_n)^\top$ being uniformly independently distributed in the domain $X := \prod_{i=1}^{n}[a_i, b_i]$. The output z is defined by the vector $\mathbf{C}$, which is often independent of physical parameters. The dynamical system (15.1) consists of ordinary differential equations (ODEs) for a regular matrix $\mathbf{E}$ and of differential algebraic equations (DAEs) for a singular matrix $\mathbf{E}$, respectively. Furthermore, we assume that the system (15.1) is asymptotically stable for all $\mathbf{x} \in X$.

The input-output behavior of the dynamical system (15.1) in the frequency domain is described by a transfer function, see [1, Ch. 4]. More precisely, let U, Z be the Laplace transforms of the input u and the output z, respectively. It follow that

$$Z(s, \mathbf{x}) = H(s, \mathbf{x})U(s)$$

for $s \in S(\mathbf{x}) \subseteq \mathbb{C}$ and each $\mathbf{x} \in X$ with the transfer function

$$H(s, \mathbf{x}) = \mathbf{C}(\mathbf{x})(s\mathbf{E}(\mathbf{x}) - \mathbf{A}(\mathbf{x}))^{-1}\mathbf{B}(\mathbf{x}) \in \mathbb{C}. \qquad (15.2)$$

Due to the stability property of (15.1) the transfer function (15.2) is defined on the imaginary axis.

15.2.2 Derivative-Based Measures

The output of (15.1) can be standardized to a function $f(\mathbf{x})$ defined in the unit hypercube $\mathcal{H}^n := [0, 1]^n$ in the time domain as well as the frequency domain. Specifically, the entities underlying the function $f(\mathbf{x})$ are the temporal evolution of z in (15.1) and the real-valued system responses in the frequency domain, i.e., magnitude and phase.

Assuming that f is differentiable, functionals depending on $\frac{\partial f}{\partial x_i}$ are proposed as estimators for the sensitivity with respect to x_i. The modified Morris measure [5] based on absolute values reads as

$$\mu_i = \int_{\mathcal{H}^n} \left| \frac{\partial f}{\partial x_i} \right| d\mathbf{x} \qquad \text{for } i = 1, \ldots, n. \qquad (15.3)$$

The integral of the squared derivatives yields

$$v_i = \int_{\mathcal{H}^n} \left(\frac{\partial f}{\partial x_i} \right)^2 d\mathbf{x} \qquad \text{for } i = 1, \ldots, n. \qquad (15.4)$$

In contrast to the variance-based sensitivity measures, μ_i, v_i may become arbitrarily large. However, it is evident that the Cauchy-Schwarz inequality implies $\mu_i \leq \sqrt{v_i}$. A constraint between derivative-based and variance-based sensitivities holds true, i.e.,

$$S_i^T \leq \frac{v_i}{\pi^2 D}, \qquad \text{for } i = 1, \ldots, n \qquad (15.5)$$

where D denotes the total variance of $f(\mathbf{x})$ and S_i^{T} is the total effect sensitivity index of the ith random parameter, whose definition can be found, for example, in [4, 8, 10].

15.3 Numerical Approaches

The partial derivative w.r.t. the ith parameter is approximated using a finite difference

$$\frac{\partial f}{\partial x_i} \approx D_i f := \frac{1}{\Delta x_i} (f(\mathbf{x} + \Delta_i \mathbf{x}) - f(\mathbf{x}))$$

with $\Delta_i \mathbf{x} = (0, \ldots, 0, \Delta x_i, 0, \ldots, 0)^{\mathrm{T}}$. Typically, a small stepsize $\Delta x_i > 0$ is chosen in dependence on the machine precision ϵ_0. We apply $\Delta x_i = \sqrt{\epsilon_0} \max\{10^{-2}, |x_i|\}$. Let a quadrature formula or a sampling technique be given with the nodes $\mathbf{x}^j \in \mathcal{H}^n$ and the weights $w_j \in \mathbb{R}$ for $j = 1, \ldots, N$. The discretization of the measure (15.3) has the form

$$\tilde{\mu}_i := \sum_{j=1}^{N} \frac{w_j}{\Delta x_i} \left| f\left(\mathbf{x}^j + \Delta_i \mathbf{x}^j\right) - f\left(\mathbf{x}^j\right) \right| \tag{15.6}$$

and of the measure (15.4)

$$\tilde{v}_i := \sum_{j=1}^{N} \frac{w_j}{(\Delta x_i)^2} \left(f\left(\mathbf{x}^j + \Delta_i \mathbf{x}^j\right) - f\left(\mathbf{x}^j\right)\right)^2. \tag{15.7}$$

The computational cost of (15.6) as well as (15.7) is characterized by $(n + 1)N$ function evaluations of f.

Sampling utilizing QMC yields $w_j = \frac{1}{N}$ for all $j = 1, \ldots, N$. This approach generates a sequence of low discrepancy in the unit hypercube (see [7]). Note that a disadvantage of QMC is a relatively slow convergence rate of about $\mathcal{O}(N^{-1})$.

Alternatively we consider cubature rules with polynomial exactness of degree three. Cubature rules with higher degree of polynomial exactness exhibit a number of nodes, which grows quadratically with the dimension. Thus the number of nodes becomes too large. In this case the Stroud-3 approach (see [9, 12]) is proposed since it is the unique rule with polynomial exactness of degree three and a minimum number of nodes, i.e., $N = 2n$ (see [2]). The Stroud-3 method is applicable to our problem, because the integrands are sufficiently smooth. More precisely, the Stroud-3 approach generates nodes

$$x_{2k-1}^j = \sqrt{\frac{2}{3}} \cos\left(\frac{(2k-1)j\pi}{n}\right), \qquad x_{2k}^j = \sqrt{\frac{2}{3}} \sin\left(\frac{(2k-1)j\pi}{n}\right)$$

in $[-1, 1]^n$ for the components $k = 1, 2, \ldots, \lfloor \frac{n}{2} \rfloor$ and if n is odd $x_n^j = \frac{(-1)^j}{\sqrt{3}}$. After a linear transformation mapping the nodes from $[-1, 1]^n$ onto $\mathcal{H}^n$ the weights of the cubature all become $w_j = \frac{1}{2n}$. Unfortunately the accuracy of the Stroud-3 method cannot be improved any further because the number of nodes is fixed. For this reason the only way for increasing the accuracy seems to carry out a Stroud formula of a higher order where the nodes cannot be propagated for increasing the accuracy. Therefore we are not able to reuse the function evaluations because the grids for different orders are not nested. Inversely, in QMC the sequences of the function evaluations can be continued as long as desired. A further disadvantage of the cubature approach is that an adaptive method for an error control based on estimates is not available directly.

15.4 Application Example

We consider a linear dynamical system of the form (15.1), which represents the mathematical model of a band-stop filter, see [11, p. 350]. Figure 15.1 depicts this band-stop circuit consisting of a chain of K cells.

As random physical parameters, we obtain K capacitances (group I) parallel to K inductances (group I) and $K - 1$ capacitances (group II) in line with $K - 1$ inductances (group II). The resistance at the input and the load resistance at the output are kept constant. A single input voltage and a single output voltage appear. Then the dynamical system results in a linear SISO type. Using modified nodal analysis [3] a system of DAEs with nilpotency index one models the transient behavior of all node voltages and some branch currents.

For the numerical experiments a problem size of $K = 26$ cells is selected. The total number of capacitances and inductances results to $n = 4K - 2 = 102$. We replace these parameters by random variables with independent uniform distributions having ranges of 10% around their mean values. The numbers of used function evaluations (NOFE) for computing the derivative-based sensitivity measures are listed in Table 15.1.

We compute the transfer function on the imaginary axis in the frequency window $\omega \in [10^{-1}, 10^1]$. Figure 15.2 depicts the expected value and the total variance for magnitude and phase obtained by Stroud-3 which implicates 204 nodes and

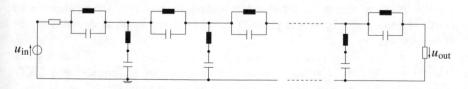

Fig. 15.1 Band-stop filter

Table 15.1 Different numerical approaches

Methods	No. points	NOFE
Stroud-3	$n = 102, N = 2n = 204$	21,012
QMC	$N = 1000$	103,000

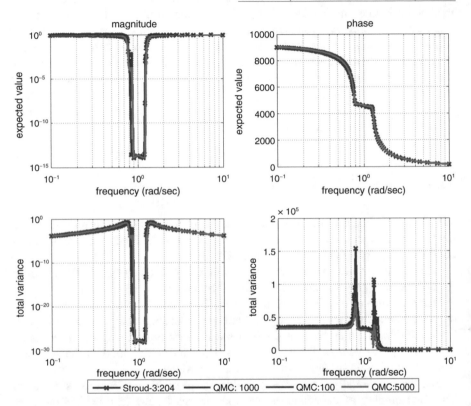

Fig. 15.2 Expected value and total variance of magnitude and phase

QMC based on a Sobol sequence [7] of 100, 1000 and 5000 points, respectively. We observe a good agreement while Stroud-3 requires a lower NOFE.

Figures 15.3 and 15.4 depict the computed derivative-based sensitivity coefficients (15.3) using Stroud-3 and QMC, respectively. Both amplitude and phase exhibit a good agreement. Comparing with variance-based sensitivities, see Fig. 15.5, which are computed by a double-sum approach of Stroud-3, cf. [4], we observe that these two types of sensitivity measurements have a strong consistency in the detection of significant parameters as well as sensitive domains in the frequency window.

For analyzing the accuracy we examine the constraint (15.5). The differences $\pi^2 S_i^T D - v_i$ for all $i = 1, \ldots, n$ are implemented using Stroud-3 and QMC, respectively. Here we compute S_i^T using the single-sum approach subject to QMC, cf. [4]. Figure 15.6 depicts $\max_i(\pi^2 S_i^T D - v_i)$ for the individual parameter groups.

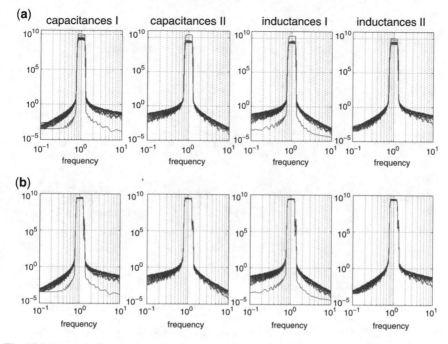

Fig. 15.3 Numerical results for derivative-based sensitivity coefficients μ_i of amplitude computed by (**a**) Stroud-3; (**b**) QMC

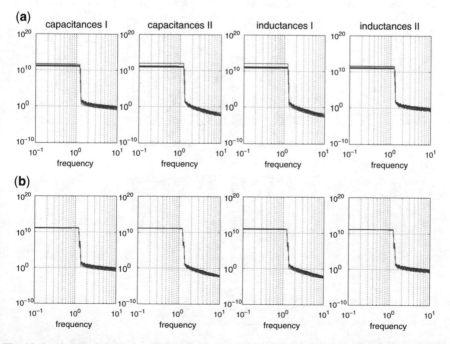

Fig. 15.4 Numerical results for derivative-based sensitivity coefficients μ_i of phase computed by (**a**) Stroud-3; (**b**) QMC

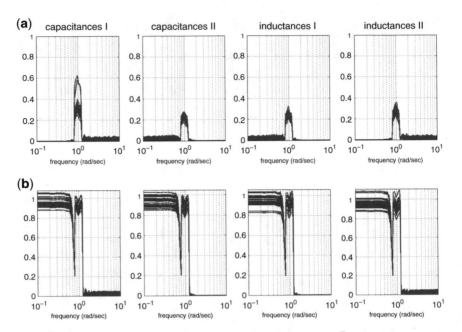

Fig. 15.5 Numerical results for variance-based sensitivity coefficients S_i^T computed by Stroud-3: (**a**) amplitude; (**b**) phase

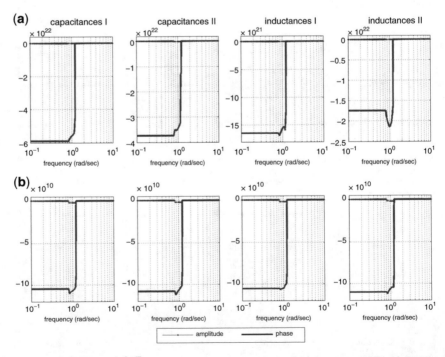

Fig. 15.6 Maximum of $\pi^2 S_i^T D - \nu_i$ for respective parameter groups: (**a**) Stroud-3; (**b**) QMC

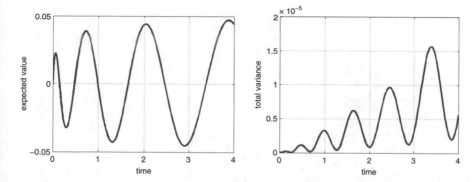

Fig. 15.7 Expected value (left) and total variance (right) of the output signal in transient simulation

The numerical solutions satisfy the inequality (15.5) strictly in the frequency domain.

We also simulate the system response in the time domain with the input signal $u(t) = \sin(t)$ as an instance, i.e., the input has the frequency $\omega = 1$ (rad/s). Initial values at $t = 0$ are all zeros. The expected value and the variance of the output are computed by Stroud-3 and presented in Fig. 15.7.

With the attempt to retrieve the sensitive characteristic from the frequency domain within the time window the derivative-based sensitivity characteristics in temporal evolution are computed by Stroud-3. Figure 15.8 shows a comparison between the time window and the associated frequency domain (zoom from Fig. 15.3). Obviously, a parameter dominates in the time domain if and only if it owns also a substantial influence in the associated frequency interval.

15.5   Conclusions

We conclude that both QMC and Stroud-3 are feasible to extract the significant parameters in a high-dimensional system. Furthermore, they have also successfully detected frequency intervals in which the output is especially influenced by random parameters. The Stroud-3 approach is more efficient due to a less NOFE. Investigations on accuracy issues for certain specified tolerances remain a challenge.

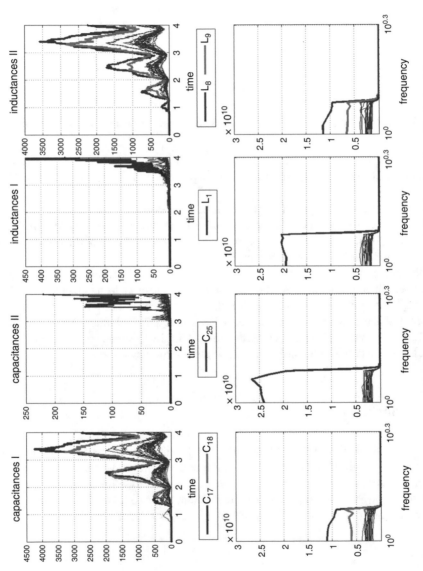

Fig. 15.8 Derivative-based sensitivity measures computed by the Stroud-3 method in the global time domain and in the local frequency domain at rate $\omega = 1$ of the input signal

References

1. Antoulas, A.C.: Approximation of Large-Scale Dynamical Systems. SIAM, Philadelphia (2005)
2. Davis, P.J., Rabinowitz, P.: Methods of Numerical Integration, 2nd edn. Academic, London (1984)
3. Ho, C.-W., Ruehli, A., Brennan, P.: The modified nodal approach to network analysis. IEEE Trans. Circuits Syst. 22(6), 504–509 (1975)
4. Liu, Q., Pulch, R.: A comparison of global sensitivity analysis methods for linear dynamical systems. Research report, University of Greifswald (2016)
5. Morris, M.D.: Factorial sampling plans for preliminary computational experiments. Technometrics 33, 161–174 (1991)
6. Pulch, R., ter Maten, E.J.W., Augustin, F.: Sensitivity analysis and model order reduction for random linear dynamical systems. Math. Comput. Simul. 111, 80–95 (2015)
7. Sobol, I.M.: Distribution of points in a cube and approximate evaluation of integrals. USSR Comput. Math. Math. Phys. 7, 86–112 (1967)
8. Sobol, I.M., Kucherenko, S.: Derivative based global sensitivity measures and their link with global sensitivity indices. Math. Comput. Simul. 79, 3009–3017 (2009)
9. Stroud, A.: Remarks on the disposition of points in numerical integration formulas. Math. Comput. Simul. 11, 257–261 (1957)
10. Sudret, B.: Global sensitivity analysis using polynomial chaos expansions. Reliab. Eng. Syst. Saf. 93, 964–979 (2008)
11. Whitaker, J.C.: The Electronics Handbook. CRC Press, Taylor & Francis Group, Boca Raton (2005)
12. Xiu, D., Hesthaven, J.S.: High order collocation methods for differential equations with random inputs. SIAM J. Sci. Comput. 27(3), 1118–1139 (2005)

Chapter 16
Fitting Generalized Gaussian Distributions for Process Capability Index

Theo G. J. Beelen, Jos J. Dohmen, E. Jan W. ter Maten, and Bratislav Tasić

Abstract The design process of integrated circuits (IC) aims at a high yield as well as a good IC-performance. The distribution of measured output variables will not be standard Gaussian anymore. In fact, the corresponding probability density function has a more flat shape than in case of standard Gaussian. In order to optimize the yield one needs a statistical model for the observed distribution. One of the promising approaches is to use the so-called Generalized Gaussian distribution function and to estimate its defining parameters. We propose a numerical fast and reliable method for computing these parameters.

16.1 Introduction

In circuit design one aims to reduce faults and to increase yield [8]. Specially added electronic control is applied to obtain narrow tails in empirical probability density functions. This process is called (electronic) 'trimming'. It has no relation to statistical techniques like Winsoring (in which one clips outliers to a boundary percentile), or Trimming (in which one simply neglects outliers). Here it is an electronic tuning, f.i., by a variable resistor. Assume that at some measurement point a circuit has a DC solution $V(R, p)$, that depends on a resistor R and an uncertain parameter p. The circuit design aims to satisfy a performance criterion $V_{\text{Low}} \leq$

T. G. J. Beelen
Eindhoven University of Technology, Eindhoven, The Netherlands
e-mail: T.G.J.Beelen@tue.nl

J. J. Dohmen · B. Tasić
NXP Semiconductors, Eindhoven, The Netherlands
e-mail: Jos.J.Dohmen@nxp.com; Bratislav.Tasic@nxp.com

E. J. W. ter Maten (✉)
Eindhoven University of Technology, Eindhoven, The Netherlands

Bergische Universität Wuppertal, Wuppertal, Germany
e-mail: E.J.W.ter.Maten@tue.nl; Jan.ter.Maten@math.uni-wuppertal.de;
termaten@math.uni-wuppertal.de

© Springer International Publishing AG, part of Springer Nature 2018
U. Langer et al. (eds.), *Scientific Computing in Electrical Engineering*,
Mathematics in Industry 28, https://doi.org/10.1007/978-3-319-75538-0_16

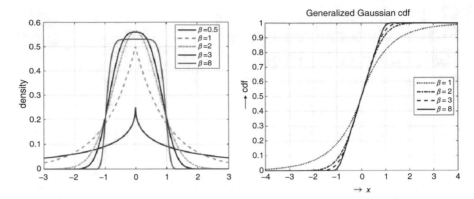

Fig. 16.1 Left: Generalized Gaussian probability density function $f(x)$ as in (16.1) with $\mu = 0$ and $\alpha = 1$. Right: Generalized Gaussian cumulative density function $F(x)$ as in (16.2)

$V \leq V_{\text{Up}}$. Now for each p we may determine how V depends on R. An optimal $R(p)$ assures that $V(R(p), p) = V_{\text{Ref}} \in [V_{\text{Low}}, V_{\text{Up}}]$, but there is no guarantee that this can be achieved for all p. If $R(p)$ exists, $R(p)$ can be determined by some nonlinear solution technique, involving solving the circuit equations several times. More general, we determine $R(p)$ such that $|V(R(p), p) - v|$ for $v \in [V_{\text{Low}}, V_{\text{Up}}]$ is minimum. The effect is that a probability density function (pdf) becomes more concentrated around a mean value and tails become more narrow.

We assume N independent samples x_i are obtained from an electronic trimming process in some given interval $[A, B]$, and based on some empirical density function. To define a quality measure index (in Sect. 16.3) we are first interested in the 'best' fitting function within the family of Generalized Gaussian Density (GGD) distributions as shown in Fig. 16.1 and given by the expression

$$f(x) = \frac{\beta}{2\alpha \, \Gamma(1/\beta)} \exp\left(-\left(\frac{|x - \mu|}{\alpha}\right)^{\beta}\right),\tag{16.1}$$

where $\alpha, \beta > 0$, $\mu \in \mathbb{R}$ and $\Gamma(z) = \int_0^\infty t^{z-1} e^{-t} dt$, for $z > 0$, is the Gamma function [7, 10]. The mean and the variance of the GGD (16.1) are given by μ and $\alpha^2 \Gamma(3/\beta)/\Gamma(1/\beta)$, respectively. Hence after expressing $\alpha = \sigma \sqrt{\Gamma(1/\beta)/\Gamma(3/\beta)}$ we get that, for all β, the variance is σ^2.

Notice that if $x \leq \mu$ then the cumulative distribution function (cdf) $F(x)$ corresponding to the GGD (16.1) is given by

$$F(x) = \frac{\beta}{2\alpha\Gamma(1/\beta)} \int_{-\infty}^{x} \exp\left(-\left(\frac{|y - \mu|}{\alpha}\right)^{\beta}\right) dy$$

$$= \frac{1}{2\Gamma(1/\beta)} \int_{((\mu-x)/\alpha)^{\beta}}^{\infty} z^{(1/\beta)-1} \exp(-z) dz.\tag{16.2}$$

A graphical impression of $f(x)$ and $F(x)$ are given in Fig. 16.1.

By using the Complementary Incomplete Gamma function defined by

$$\Gamma(a, x) = \int_x^\infty t^{a-1} exp(-t)dt \tag{16.3}$$

we can rewrite (16.2) as

$$F(x) = \frac{\Gamma\left(1/\beta, \left(\frac{\mu-x}{\alpha}\right)^\beta\right)}{2\Gamma(1/\beta)} \tag{16.4}$$

This can be further simplified using the Upper Incomplete Gamma function [9, 10] for which standard software is available. We note that for $\beta = 2$ one has $\Gamma(1/2) = \sqrt{\pi}$, $\Gamma(3/2) = 0.5\sqrt{\pi}$ and then $\alpha = \sigma\sqrt{2}$; i.e., the GGD becomes the Gaussian distribution. The parameter β determines the shape. For $\beta = 1$ the GGD corresponds to a Laplacian distribution; for $\beta \to +\infty$ the probability density function (pdf) in (16.1) converges to a uniform distribution in $(\mu - \sqrt{3}\sigma, \mu + \sqrt{3}\sigma)$, and when $\beta \downarrow 0$ we get a degenerate distribution in $x = \mu$ (but with a finite variance). For some graphical impression, see Fig. 16.1.

In our applications we are facing broad pdfs with relatively long, but steep tails and we want to estimate them accurately, so we are interested in the cases when $\beta \geq 2$.

Despite the fact that several distributions of output results will not be symmetrical, we restrict ourselves here to the family of GGD (16.1).

The parameters of the 'best' fitting distribution function can be found by maximizing the logarithm of the likelihood function $L = \ln(\mathscr{L}) = \sum_{i=1}^N f(x_i)$. The necessary conditions are

$$\frac{\partial L}{\partial \alpha} = 0 : \alpha = \left(\frac{\beta}{N}\sum_{i=1}^N |x_i - \mu|^\beta\right)^{1/\beta}, \tag{16.5}$$

$$\frac{\partial L}{\partial \beta} = 0 : \frac{1}{\beta} + \frac{\Psi(1/\beta)}{\beta^2} - \frac{1}{N}\sum_{i=1}^N \left|\frac{x_i - \mu}{\alpha}\right|^\beta \ln\left|\frac{x_i - \mu}{\alpha}\right| = 0, \tag{16.6}$$

$$\frac{\partial L}{\partial \mu} = 0 : \sum_{x_i \geq \mu} |x_i - \mu|^{\beta-1} - \sum_{x_i < \mu} |x_i - \mu|^{\beta-1} = 0. \tag{16.7}$$

Here Ψ is the Digamma function $\Psi(x) = \frac{d}{dx}\ln(\Gamma(x)) = \Gamma'(x)/\Gamma(x)$, see [10]. Let us first assume that $\mu = \hat{\mu}$ is known or well estimated. Then we can ignore (16.7). Several papers [2–4] consider estimates for α and β to solve the Eqs. (16.5)–(16.6), but they assume that the sample size is large enough and/or that $\beta \leq 3$, motivated by the various application areas. We note that [5, 6] also consider the case for a small sample size.

We exploit the explicit elimination of α in (16.6) after which only one additional equation remains

$$g(\beta) = g(\beta; \hat{\mu}) = 0 \qquad (16.8)$$

in which $\mu = \hat{\mu}$ is now a given parameter. The analytical formulae for $g(\beta)$ and $g'(\beta)$ are given by (see also [1, 4, 6, 10])

$$g(\beta) = 1 + \frac{\Psi(1/\beta)}{\beta} - \frac{\sum_{i=1}^{N} |x_i - \hat{\mu}|^\beta \ln |x_i - \hat{\mu}|}{\sum_{i=1}^{N} |x_i - \hat{\mu}|^\beta} + \frac{\ln\left(\frac{\beta}{N}\sum_{i=1}^{N} |x_i - \hat{\mu}|^\beta\right)}{\beta},$$

$$g'(\beta) = -\frac{\Psi(1/\beta)}{\beta^2} - \frac{\Psi'(1/\beta)}{\beta^3} + \frac{1}{\beta^2}$$

$$-\frac{\sum_{i=1}^{N} |x_i - \hat{\mu}|^\beta \left(\ln |x_i - \hat{\mu}|\right)^2}{\sum_{i=1}^{N} |x_i - \hat{\mu}|^\beta} + \left(\frac{\sum_{i=1}^{N} |x_i - \hat{\mu}|^\beta \ln |x_i - \hat{\mu}|}{\sum_{i=1}^{N} |x_i - \hat{\mu}|^\beta}\right)^2$$

$$+\frac{\sum_{i=1}^{N} |x_i - \hat{\mu}|^\beta \ln |x_i - \hat{\mu}|}{\beta \sum_{i=1}^{N} |x_i - \hat{\mu}|^\beta} - \frac{\ln\left(\frac{\beta}{N}\sum_{i=1}^{N} |x_i - \hat{\mu}|^\beta\right)}{\beta^2}.$$

$$(16.9)$$

Clearly, (16.8) can be solved by any (nonlinear) iterative method, for example by Newton's method using the expressions in (16.9). We outline our algorithm in Algorithm 1.

If μ is not a priori known then we can apply an iterative process with an estimator $\hat{\mu}$ for μ and solving (16.8) for β, giving an estimator $\hat{\beta}$ that can be used for finding $\hat{\alpha}$ with (16.5). If (16.8) with $\hat{\mu}$ can not be solved with sufficient accuracy, then the process is repeated with a new estimator for μ.

Algorithm 1: Averaged generalized Gaussian distribution fit

1: **procedure** AGGDF ($\mathbf{X}$, N, M, $\hat{\mu}$)
2:       Determine the empirical pdf $\hat{f}(x)$ from the data $\mathbf{X}$.   ▷ See Fig. 16.2
3:       Compute the cumulative distribution function $\hat{F}(x) = \int_{-\infty}^{x} \hat{f}(t)\mathrm{d}t$.
4:       **for** $k = 1, \ldots, M$ **do**
5:             Generate random values $\{x_i^k \mid i = 1, \ldots, N\}$ using $\hat{F}^{-1}$.
6:             Compute the zero $\hat{\beta}_k$ of $g(\beta) = 0$, using these x_i^k-values and $\hat{\mu}$. ▷ See (16.8)
7:             Compute $\hat{\alpha}_k$.   ▷ See (16.5)
8:       **end for**
9:       Average $\hat{\beta} = \frac{1}{M}\sum_{i=1}^{M} \hat{\beta}_k$, $\hat{\alpha} = \frac{1}{M}\sum_{i=1}^{M} \hat{\alpha}_k$.
10:      **return** $\hat{\alpha}, \hat{\beta}$.   ▷ In this Algorithm $\hat{\mu}$ is unchanged
11: **end procedure**

After having computed the parameters $\hat{\mu}, \hat{\alpha}, \hat{\beta}$ we consider the resulting density function $f(x; \ \hat{\mu}, \hat{\alpha}, \hat{\beta})$ as best fit to the measured data (see [1]). We make the following observation [1]. We introduced M-times the steps 5–7 within a loop and taking averages, see Algorithm 1. One can choose $M = 1$ and N sufficiently large (usually $N \gg 1000$). In our case, $M = 50$, $N = 200$. So, N can be taken smaller.

In our numerical experiments we observed that due to the large value of $\partial \alpha / \partial \beta$ averaging the $\hat{\alpha}_k$ gives better results for α than by using (16.5) on $\hat{\beta}$.

16.2 Numerical Results

We applied Algorithm 1 (with $M = 50$ and $N = 200$) to 'trimmed' data from first NXP IC-measurements (Fig. 16.2). The computed values β_k and their mean $\hat{\beta}$ are shown in Fig. 16.3-(left). The computed density function f as well as the initially fitted (non-symmetrical) density function $\hat{f}$ are given in Fig. 16.4. Note that the tails are very well approximated in Fig. 16.3-(right). To get an impression of the sensitivity of the computed density w.r.t. $\hat{\alpha}$ we varied the computed value of $\hat{\alpha}$ with $\pm 10\%$, plotted the corresponding densities and computed the Mean Square Error (MSE). See Fig. 16.4 and [1]. Notice that in Fig. 16.4 a 10% variation in α has a large effect on the pdf f. Clearly, the approximation of f around its top using $\hat{\alpha}$ is better than the pdfs with $\hat{\alpha} \pm 10\%$. A similar observation holds for the slopes of pdf f. The best fit $\hat{\alpha}$ was obtained by the mean as in step 9 in Algorithm 1. So, we consider $\hat{\alpha}$ as best fit.

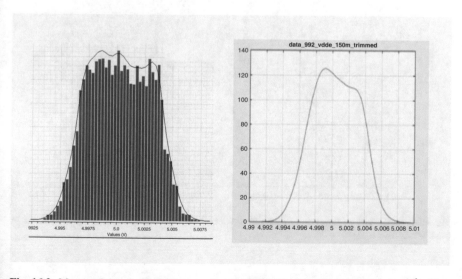

Fig. 16.2 Measured data (left) and the associated empirical probability density function $\hat{f}$ (right)

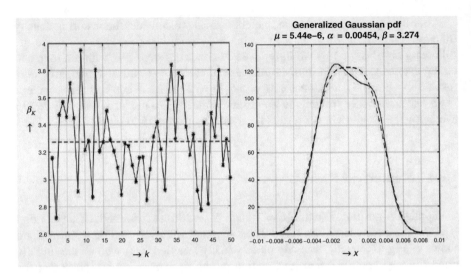

Fig. 16.3 (left): The computed β_k with mean $\hat{\beta} = 3.27$ and $|\beta_k - \hat{\beta}| < 20\%$. (right): The empirical probability function (solid) and the final fitted GGD (dashed)

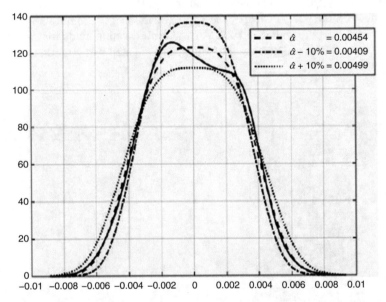

Fig. 16.4 Sensitivity of the density function f w.r.t. $\hat{\alpha}$. $MSE = (14.31, 56.94, 91.95)$ for $\hat{\alpha} = (454, 499, 409) \times 10^{-5}$

16.3   A Quality Measure Index for a Generalized Gaussian Distribution

Assuming an underlaying distribution being standard Gaussian, the capability of a manufacturing process can be measured using some process capability indices like

$$C_p = \frac{U - L}{6\sigma} \quad \text{and} \quad C_{pk} = \frac{\min (U - \mu, \mu - L)}{3\sigma}, \tag{16.10}$$

where $[L, U]$ is the specification interval, μ is the process mean and σ is the process standard deviation and a process is said to be capable if the process capability index exceeds a value $k \geq 1$, where usually $k = 4/3$. In case of a GGD (16.1) we can introduce a capability index C_{pkg} as quality indicator, similar to the standard Gaussian case as

$$C_{pkg} = \frac{\min (U - \mu, \mu - L)}{3\sigma}, \tag{16.11}$$

where $\sigma^2 = \alpha^2 \Gamma(3/\beta)/\Gamma(1/\beta)$.

L and U are the lower and upper tolerance levels, respectively.

Hence, the C_{pkg} simply results as a post processing facility of Algorithm 1, after having determined α, β and μ. In practice, formula (16.11) can be applied in two ways. First, a specific value $C_{pkg} = c$ can be given in order to meet certain yield requirements. Assuming $|\mu - L| < |U - \mu|$, then L can be computed via $c = C_{pkg} = (\mu - L)/\sigma$. On the other hand, if L and U are known from product specs, then C_{pkg} can be determined as a measure for the yield.

16.4   Conclusions

We have shown that measured IC chip production data can adequately be modelled by a Generalized Gaussian distribution (GGD). We developed a new robust numerical procedure for computing the parameters of such GGD. The GGD did fit very accurately. Using the GGD a quality measure can be defined analogously to the CPK index for standard Gaussian distributions.

Acknowledgements The authors acknowledge support from the projects **CORTIF** (http://cortif. xlim.fr/): *Coexistence Of Radiofrequency Transmission In the Future*, a CATRENE project (Cluster for Application and Technology Research in Europe on NanoElectronics, http://www.catrene.org/) and **nanoCOPS** (http://fp7-nanocops.eu/): *Nanoelectronic COupled Problems Solutions*, FP7-ICT-2013-11/619166.

References

1. Beelen, T.G.J., Dohmen, J.J.: Parameter estimation for a generalized Gaussian distribution. CASA Report 15-40, TU Eindhoven (2015). http://www.win.tue.nl/analysis/reports/rana15-40. pdf
2. Bombrun, L., Pascal, F., Tourneret, J.-Y., Berthoumieu, Y.: Performance of the maximum likelihood estimators for the parameters of multivariate generalized Gaussian distributions. In: Proceedings of ICASSP-2012, IEEE. International Conference on Acoustics, Speech, and Signal Processing, Kyoto, pp. 3525–3528 (2012). https://hal.archives-ouvertes.fr/hal-00744600
3. González-Farías, G., Domínguez-Molina, J.A., Rodríguez-Dagnino, R.M.: Efficiency of the approximated shape parameter estimator in the generalized Gaussian distribution. IEEE Trans. Veh. Technol. **58**(8), 4214–4223 (2009)
4. Kokkinakis, K., Nandi, A.K.: Exponent parameter estimation for generalized Gaussian probability density functions with application to speech modeling. Signal Process. **85**, 1852–1858 (2005)
5. Krupiński, R.: Approximated fast estimator for the shape parameter of generalized Gaussian distribution for a small sample size. Bull. Pol. Acad. Sci. Tech. Sci. **63**(2), 405–411 (2015)
6. Krupiński, R., Purczyński, J.: Approximated fast estimator for the shape parameter of generalized Gaussian distribution. Signal Process. **86**, 205–211 (2006)
7. Martinez, W.L., Martinez, A.R.: Computational Statistics Handbook with Matlab. Chapman & Hall/CRC, London (2002)
8. ter Maten, E.J.W., Wittich, O., Di Bucchianico, A., Doorn, T.S., Beelen, T.G.J.: Importance sampling for determining SRAM yield and optimization with statistical constraint. In: Michielsen, B., Poirier, J.-R. (eds.): Scientific Computing in Electrical Engineering SCEE 2010. Series Mathematics in Industry, vol. 16, pp. 39–48. Springer, Berlin (2012)
9. Temme, N.M.: Computational aspects of incomplete gamma functions with large complex parameters. Int. Ser. Numer. Math. **119**, 551–562 (1994)
10. Wikipedia: Generalized Normal Distribution. https://en.wikipedia.org/wiki/Generalized_ normal_distribution; Gamma function. https://en.wikipedia.org/wiki/Gamma_function; Digamma function. https://en.wikipedia.org/wiki/Digamma_function; Incomplete Gamma Function. https://en.wikipedia.org/wiki/Incomplete_gamma_function (2016)

Chapter 17
Robust Optimization of an RFIC Isolation Problem Under Uncertainties

Piotr Putek, Rick Janssen, Jan Niehof, E. Jan W. ter Maten, Roland Pulch, Michael Günther, and Bratislav Tasić

Abstract Modern electronics systems involved in communication and identification impose demanding constraints on both reliability and robustness of components. On the one hand, it results from the influence of manufacturing tolerances within the continuous down-scaling process into the output characteristics of electronic devices. On the other hand, the increasing integration process of various systems on a single die force a circuit designer to make some trade-offs in preventing interference issues and in compensating coupling effects. Thus, constraints in terms of statistical moments have come in a natural way into optimization formulations of electronics products under uncertainties. Therefore, in this paper, for the careful assessment of the propagation of uncertainties through a model of a device a type of Stochastic Collocation Method (SCM) with Polynomial Chaos (PC) was used. In this way a response surface model can be included in a stochastic, constrained optimization problem. We have illustrated our methodology on a Radio Frequency Integrated Circuit (RFIC) isolation problem. Achieved results for the optimization confirmed efficiency and robustness of the proposed methodology.

P. Putek (✉) · E. J. W. ter Maten · M. Günther
Bergische Universität Wuppertal, Wuppertal, Germany
e-mail: putek@math.uni-wuppertal.de; termaten@math.uni-wuppertal.de;
guenther@math.uni-wuppertal.de

R. Janssen · J. Niehof · B. Tasić
NXP Semiconductors, Eindhoven, The Netherlands
e-mail: Rick.Janssen@nxp.com; Jan.Niehof@nxp.com; Bratislav.Tasic@nxp.com

R. Pulch
Ernst-Moritz-Arndt-Universität Greifswald, Institute of Mathematics and Computer Science,
Greifswald, Germany
e-mail: roland.pulch@uni-greifswald.de

© Springer International Publishing AG, part of Springer Nature 2018
U. Langer et al. (eds.), *Scientific Computing in Electrical Engineering*,
Mathematics in Industry 28, https://doi.org/10.1007/978-3-319-75538-0_17

17.1   Introduction

Due to the continuous advances in semiconductor technology, modern mixed-signal and radio frequency (RF) integrated circuits (ICs) show a tendency to increase the integration of various systems on a singe die [10]. This trend in electronics results not only in decreasing material cost but also allows for easier implementation of multiple functions in a compact unit [8]. On the one hand, this complexity gives challenges in the integration of noisy parts, the so-called aggressors, as well as sensitive parts, the so-called victims, and other intellectual property blocks (IPs) to provide its proper and interference-free functioning. On the other hand, the integration process has also impact on the failure probability of nanoscale or molecular scale devices associated with yield loss, which can be caused by defects, faults, process variations and design issues [15]. In this respect, the impact of statistical variations in input parameters onto the output characteristics of electronic devices has played an increasingly important role in the predictability and reliability of simulations. Actually, these statistical variations, resulting from manufacturing tolerances of industrial processes, could lead to the acceleration of migration phenomena in semiconductor devices and finally can cause a thermal destruction of devices due to thermal runaway [12, 14]. Moreover, unintended RF coupling, which can occur both as a result of industrial imperfections and as a consequence of the integration process, might additionally downgrade the quality of products and their performance or even be dangerous for safety of both environment and the end users [5]. It should be pointed out, though, that meeting the specification requirements for electromagnetic compatibility standards [1] and issues related to interference between IPs at early design stages allows for avoiding expensive re-spins and for the consecutive decrease of the time-to-market cycle. The ICs designer needs to take special attention to interference issues during all the stages of the product development cycle. Therefore, a structured approach to find an optimal isolation configuration of the IC design needs to be applied.

Our new contribution relies first on incorporating the uncertainty quantification (UQ) analysis into the modeling of electronic devices to provide reliable and robust simulations. In work [5] some theoretical foundations can be found. Thus, in our work we address the stochastic optimization problem, described by the system of the stochastic differential algebraic equations (SDAEs). Next, the optimization procedure for the compensation of the aggressor impact on the proper operation of the IC system is proposed and is successfully applied. In this context, the insist is given more on a new application. However, in the contrast to our previous work [12, 14], where we considered the stochastic partial differential equations (SPDEs) coupled optimization problem using the topological derivative method, here we deal with the SDAEs model, where the regularized Gauss-Newton algorithm has been used [11]. Specifically, our approach has been tailored to the investigation of the coupling path via an exposed diepad, downbonds and bondwires in order to find their optimal configuration, which ensures the minimal influence of the digital noise on the device functioning under uncertainties. As to the best knowledge of authors, this important engineering problem is investigated for the first time not only in the robust but also in the deterministic optimization framework. To this end, we incorporated our automated optimization procedure in the flow of the floorplanning and grounding strategy [10].

17.2   Modeling Approach

In our research, we consider an integrated RFCMOS automotive transceiver design as a case test, which is shown in Fig. 17.1. This is a fully functional chip that consists of four main domains including (a) the analog-to-digital converter (ADC), (b) the receiver and power amplifier (RxPA), (c) the crystal oscillator and local oscillator (XOLO) and (d) the digital part. The latter is responsible for a noise generation, which disturbs the other mentioned subsystems. For the simulation of a chip architecture, the software ADS/Momentum from Keysight Technologies has been used [9], which employs the Methods of Moments (MoM) [3]. Therein, the concept of Green functions is used to model the proper behavior of the substrate [7].

17.2.1   Field Model of the Integrated Circuit

In order to take the interaction of the ICs with their physical environment into account, an integral formulation of the time-harmonic PDEs, derived from

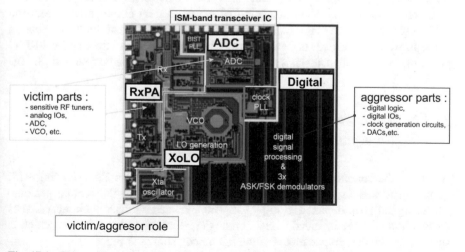

Fig. 17.1   Chip architecture with indicated domain setup [10]

Maxwell's equations

$$\begin{cases} \nabla \cdot \epsilon\,(\theta)\,[\nabla V\,(\theta) + \mathrm{i}\,\omega\,\mathbf{A}\,(\theta)\,] = \rho\,(\theta) \\ \nu\,(\theta)\,\nabla \times \nabla \times \mathbf{A}\,(\theta) = \mathbf{J}\,(\theta) + \omega^2\,\epsilon\,(\theta)\left[\mathbf{A}\,(\theta) - \mathrm{i}\,\frac{\nabla V(\theta)}{\omega}\right] \\ \nabla \cdot \mathbf{A}\,(\theta) + \mathrm{i}\omega k V\,(\theta) = 0 \\ \nabla \cdot \mathbf{J}\,(\theta) + \mathrm{i}\omega\rho\,(\theta) = 0, \end{cases} \qquad (17.1)$$

endowed with suitable initial and boundary conditions, was solved in the Momentum. Here, V is the electric scalar potential, $\mathbf{A}$ denotes the magnetic vector potential and $\theta := (\mathbf{r}, f)$, where $\mathbf{r}$ is the location in space and $\omega = 2\pi f$ is the angular velocity with f the frequency. Furthermore, σ, ϵ and ν are real functions of space, which describe the electric conductivity, the permittivity and the reluctivity. The domain $\mathbb{R}^3 \supset D = D_1 \cup D_2 \cup D_3$ includes metal, insulator and semiconductor regions. The charge density ρ is defined as $\rho = q\,(n - p - N_\mathrm{D})$ on D_3 and 0 on $(D_{1,2})$; the current density $\mathbf{J}$ is described by $\mathbf{J}_{D_1} = -\sigma\,(\nabla V + \mathrm{i}\,\epsilon\,\omega\mathbf{A})$, $\mathbf{J}_{D_2} = 0$ and $\mathbf{J}_{D_3} = \mathbf{J}_n + \mathbf{J}_p$. Here, $\mathbf{J}_n$ and $\mathbf{J}_p$ are electron and hole current densities, while n and p represent electron and hole concentrations. N_D refers to the doping concentration, k is a constant that depends on the scaling scenario.

In our simulations, the time-harmonic analysis is applied, which provides accurate electromagnetic simulation performance at radio frequencies for the geometrically complex and electrically small designs. As output of these simulations, S-parameters can be generated for general planar circuits, which contains sufficient information to characterize each individual component. Additionally, the application of the ADS tool, allows for modeling the behavior of RF passive component by a frequency independent lumped model [9]. Hence, the lumped model can be further employed to speed up the electrical performance for an RFIC optimization problem. However, since the extraction of the equivalent circuit model from the PDEs equation is not a major topic of this paper, we briefly recall only some basics concept, which is needed for incorporating the lumped model into an robust RFIC problem. More detailed analysis subjected the former topic can be found in [8, 10].

17.2.2 Equivalent Circuit Model and Floorplanning/Grounding Strategy

Based on the conducted simulations of (17.1) and the proper floorplanning with grounding strategies, an equivalent circuit model (EMC) of the IC and package was designed [10], depicted in Fig. 17.2. The analyzed equivalent circuit consists of the main package included the IC connections located on the left side, which is supposed to be the lumped model of the 3D model, shown in Fig. 17.1.

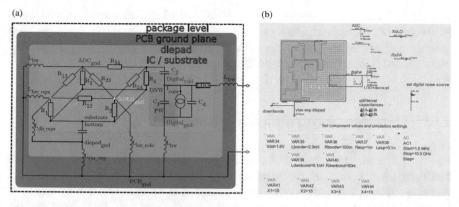

Fig. 17.2 EMC of the packaged integrated circuit. (**a**) Floorplan model for testing isolation and grounding strategies [10, 18]. (**b**) Implementation of the lumped model in ADS/Momentum

When the charge-oriented modified nodal analysis (MNA) [6] is used, the time-harmonic structured DAEs takes the form

$$i\,\omega\,\mathbf{M}\,\mathbf{q}\,(\mathbf{x}(f,\mathbf{p}),\mathbf{p}) = \mathbf{g}(\mathbf{x}(f,\mathbf{p}),\mathbf{p}), \qquad (17.2)$$

Here, $\mathbf{x}$ is a vector of unknowns, including the nodal potentials and the branch currents of inductances and voltage sources. The vector function $\mathbf{q}(\mathbf{x},\mathbf{p})$ includes the charges of capacitances and the fluxes of inductances. The vector function $\mathbf{g}(\mathbf{x},\mathbf{p})$ describes contributions from conductances/resistors and voltage differences. Capacitances, inductances and conductances/resistors may all depend on the parameters $\mathbf{p} = \mathbf{p}(\boldsymbol{\xi})$, with $\boldsymbol{\xi}$ being the random variable. $\mathbf{M}$ denotes a (singular) incidence matrix, that is specified by a network topology.

More specifically, for our purpose, the exposed diepad and downbonds has been chosen to allow for thorough analysis with respect to a number of model parameter variations including the number of downbonds, the number of ground pins, and the number of exposed diepad vias. In this way, the cross-domain transfer function $\mathbf{y}$ from the digital to the RF domain can be considered here as victims and be included in the optimization procedure as goal functions. Here, we are particularly interested in how digital influences the other subsystems. To this end, we define, for a (scalar) cross-domain coupling the transfer function, as in [2]

$$G(\omega) = \frac{|Y|}{|X|} =: |H(i\omega)|, \qquad \phi(\omega) := \phi_Y - \phi_X = \arg(H(i\omega)), \qquad (17.3)$$

when considering a complex harmonic system with a sinusoidal component of $|X|$, an angular frequency ω and a phase $\phi := \arg(X)$ as input to a linear time-invariant system and then its corresponding output as $|Y|$ and $\phi_Y := \arg(Y)$.

Within this context, the proper floorplanning and grounding methodology allows for the identification, quantification and prediction of the cross-domain coupling in the EMC model. Therein, the overall EMC model includes key elements such as the on-chip (domain regions, padring, sealring, substrate effects), the package (ground and power pins, bondwires/downbonds, exposed diepad) and the printed circuit board (ground plane, exposed diepad connections) [10].

17.3   Uncertainty Quantification Analysis

For the UQ analysis, a type of the SCM in conjunction with the PC expansion has been used. Following the methodology proposed in [17], some parameters $\mathbf{p}(\boldsymbol{\xi}) \in \varXi$ in the model (17.2) were replaced by random variables $\mathbf{p}(\boldsymbol{\xi}) = \left[p_1(\xi_{\text{db\_rxpa}}), p_2(\xi_{\text{via\_exp}}), p_3(\xi_{\text{lb\_xolo}}), p_4(\xi_{\text{lb\_rxpa}}\right] : \mathscr{A} \rightarrow \mathbb{R}$ defined on the probability triple $(\mathscr{A}, \mathscr{F}, \mathbb{P})$. Specifically, they have been shown in Fig. 17.2. Moreover, we assume that a joint probability density function $h : \varXi \rightarrow \mathbb{R}$ exists. Let y be a quadratically integrable function. Then, a response surface model of y can be obtained by a truncated series of the PC expansion, see [17],

$$y\left(f, \mathbf{p}(\boldsymbol{\xi})\right) \doteq \sum_{i=0}^{N} v_i\left(f\right)\varPhi_i\left(\mathbf{p}(\boldsymbol{\xi})\right), \tag{17.4}$$

with a priori unknown coefficient functions v_i and predetermined basis polynomials $\varPhi_i$ with the orthogonality property $\mathbb{E}\left[\varPhi_i\varPhi_j\right] = \delta_{ij}$[1] (Kronecker delta). Therein, $\mathbb{E}$ is the expected value, associated with $\mathbb{P}$. More precisely, $v_i = \mathbb{E}\left[y\varPhi_i\right]$ (component of y along $\varPhi_i$), for which we have applied a pseudo-spectral quadrature approach with the Stroud formula of order 3 [12, 14] to approximate the unknown coefficients v_i. The basic concept of this method is first to provide the solution y at each deterministic quadrature node $\mathbf{p}^{(k)}$, $k = 1, \ldots, K$, of the system (17.2). Next, the multi-dimensional quadrature rule with associated weights w_k allows for computing

$$v_i(f) \doteq \sum_{k=1}^{K} y\left(f, \mathbf{p}^{(k)}\right) \varPhi_i\left(\mathbf{p}^{(k)}\right) w_k, \tag{17.5}$$

which represents an approximation of the exact projection of $\mathbf{y}$ along $\varPhi_i$. Finally, the moments are approximated by, cf. [17],

$$\mathbb{E}\left[y\left(f, \mathbf{p}\right)\right] \doteq v_0(f), \quad \text{Var}\left[y\left(f, \mathbf{p}\right)\right] \doteq \sum_{i=1}^{N} |v_i(f)|^2 \tag{17.6}$$

---

[1]For an orthogonal system of basis polynomials a normalization can be done straightforward, e.g.,[17].

assuming $\Phi_0 = 1$. Based on the truncated PC expansion it is also possible to perform the variance-based decomposition [16], which allows for determining and ranging the most influential input parameters according to output variations, see, e.g., [13].

17.4   Robust Optimization Problem

Finally, when considering statistical moments, an optimization problem constrained by stochastic DAEs can be reformulated into the robust single objective optimization problem [14] as follows

$$
\begin{aligned}
\min_{\mathbf{p}} \quad & \mathbb{E}\left[F(\mathbf{p})\right] + \eta \sqrt{\operatorname{Var}\left[F(\mathbf{p})\right]} \\
\text{s.t.} \quad & i\,\omega\,\mathbf{M}\,\mathbf{q}\left(\mathbf{x}(f, \mathbf{p}^{(k)}), \mathbf{p}^{(k)}\right) = \mathbf{g}(\mathbf{x}(f, \mathbf{p}^{(k)}), \mathbf{p}^{(k)}), \ k = 1, \ldots, K, \\
& \mathbf{p}_{\max_\ell} \leq \mathbf{p}_\ell \leq \mathbf{p}_{\min_\ell}, \ell = 1, \ldots, P,
\end{aligned}
\tag{17.7}
$$

where $\mathbf{p}$ is a vector of optimized parameters, while $\mathbf{p}_{\max_\ell}$ and $\mathbf{p}_{\min_\ell}$ denote their box constraints, respectively. In our case, the random-dependent functional for the prescribed weight $w_i = 0.5$ reads as

$$
F(\mathbf{p}) = \sum_{i=2} w_i |y_i(\mathbf{p})|^2,
\tag{17.8}
$$

where f_i are complex-valued functions, which yield the definition of the cross-domain coupling transfer functions, see, Fig. 17.2 as follows

$$
\begin{aligned}
y_2 &= |\mathrm{CplXolo}(f)| := \frac{|\mathrm{XOLO_{gnd}}-\mathrm{PCB_{gnd}}|}{|\mathrm{Digital_{Vdd}}-\mathrm{Digital_{gnd}}|}, \\
y_3 &= |\mathrm{CplRx}(f)| \quad := \frac{|\mathrm{RxPA_{gnd}}-\mathrm{PCB_{gnd}}|}{|\mathrm{Digital_{Vdd}}-\mathrm{Digital_{gnd}}|}.
\end{aligned}
\tag{17.9}
$$

Due to the insensitivity of $y_1 = |\mathrm{CplADC}|$ w.r.t. the input variations [13], cross-domain coupling functions such as y_2 and y_3 have been chosen for the optimization purposes.

17.5   Numerical Example and Conclusions

The model, shown schematically in Fig. 17.1, has been implemented and simulated in Momentum within the frequency range from 1 MHz to 10 GHz.

An algorithm for the UQ analysis was implemented in python using the DAKOTA v.6.2 library [4]. The least squares nonlinear optimization problem has been solved in every iteration using the normal equation method and the Tikhonov regularization [11]. The deterministic values of the individual elements for the

ECM are summarized in Table 17.1, while Table 17.2 includes the mean values of variable parameters, which were considered as impedance in the form $p_j = \overline{R}_j(1 + \delta_j) + i\omega\overline{L}_j(1 + \delta_j), j = 1, \ldots, 4$ with the magnitude of perturbation $\delta_j = 0.2$. Other deterministic resistances are defined as follows $R_1 = 13.2$ [Ω], $R_2 = 13.2$ [Ω], $R_3 = 22.7$ [Ω], $R_4 = 5.6$ [Ω], $R_{12} = 77$ [Ω], $R_{13} = 332$ [Ω], $R_{23} = 217$ [Ω], $R_{24} = 96$ [Ω] and $R_{34} = 130$ [Ω].

The final result of the robust optimization has been presented in Figs. 17.3 and 17.4 and Table 17.3 shows the optimized values, found in the fourth iteration.

Table 17.1 Chosen values for the deterministic elements of the EMC model

Elements	R_{db}	L_{db}	R_{bw}	L_{bw}	C_d	C_1	C_2
Values	100.0 [mΩ]	0.1 [nH]	100.0 [mΩ]	2.0 [nH]	2.3 [nF]	0.4547 [nF]	0.2412 [nF]

Table 17.2 The mean values for the initial configuration

Elements	$\overline{R}^0_{db\_rxpa}$	$\overline{L}^0_{db\_rxpa}$	$\overline{R}^0_{via\_exp}$	$\overline{L}^0_{via\_exp}$	$\overline{R}^0_{lb\_xolo}$	$\overline{L}^0_{lb\_xolo}$	$\overline{R}^0_{lb\_rxpa}$	$\overline{L}^0_{lb\_rxpa}$
Mean values	10.0 [mΩ]	0.02 [nH]	0.1 [mΩ]	0.01 [nH]	20.0 [mΩ]	0.4 [nH]	16.7 [mΩ]	0.33 [nH]

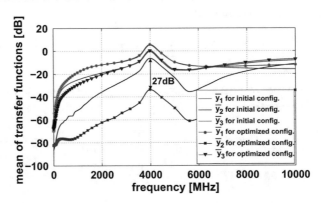

Fig. 17.3 Result for the stochastic optimization of the RFIC problem

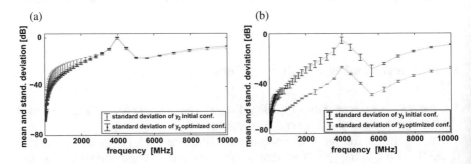

Fig. 17.4 Mean and standard deviation before and after optimization. (**a**) $y_2 = |\text{CplXolo}|$. (**b**) $y_3 = |\text{CplRx}|$

Table 17.3 The mean values for the optimized configuration

Elements	$\overline{R}_{db\_rxpa}$	$\overline{L}_{db\_rxpa}$	$\overline{R}_{via\_exp}$	$\overline{L}_{via\_exp}$	$\overline{R}_{lb\_xolo}$	$\overline{L}_{lb\_xolo}$	$\overline{R}_{lb\_rxpa}$	$\overline{L}_{lb\_rxpa}$
Mean value	9.37 [mΩ]	0.0187 [nH]	0.13 [mΩ]	0.0138 [nH]	25.0 [mΩ]	0.5 [nH]	0.36 [mΩ]	7.22 [nH]

Table 17.4 Relative error in [%] calculated for the particular functions before and after optimization

Quantities	For y_1 in [%]	For y_2 in [%]	For y_3 in [%]	For all functions in [%]
Mean value	12.41	7.64	−94.99	−24.67
Standard deviation	−90.49	−78.22	−98.77	−91.20

Both the mean values and standard deviations have been reduced significantly. However, the application of the Pareto front method instead of the average weighted method might yield the optimal solution in the sense of Pareto due to competing objective functions y_2 and y_3. This is considered as a further direction of our research. Additionally, Table 17.4 includes the information about the relative error, calculated for cross-domain functions before and after optimization.

Acknowledgements The nanoCOPS (Nanoelectronic COupled Problems Solutions) project is supported by the European Union in the FP7-ICT-2013-11 Program under the grant agreement number 619166.

References

1. Chapter SC 47A: Integrated circuits. http://www.iec.ch/emc/emc_prod/prod_main.htm
2. Chua, L.O., Lin, P.M.: Computer Aided Analysis of Electronic Circuits: Algorithms & Computational Techniques. Prentice-Hall, Englewood Cliffs (1975)
3. Collin, R.E.: Field Theory of Guided Waves. IEEE, New York (1990)
4. Dakota 6.2, Sandia National Laboratories (2015). https://dakota.sandia.gov/
5. Di Bucchianico, A., ter Maten, J., Pulch, R., Janssen, R., Niehof, J., Hanssen, J., Kapora, S.: Robust and efficient uncertainty quantification and validation of RFIC isolation. Radioengineering **23**, 308–318 (2014)
6. Feldmann, U., Günther, M.: CAD-based electric-circuit modeling in industry I: mathematical structure and index of network equations. Surv. Math. Ind. **8**(2), 97–129 (1999)
7. Gharpurey, R., Meyer, R.G.: Modeling and analysis of substrate coupling in integrated circuits. IEEE J. Solid State Circuits **31**(3), 344–353 (1996)
8. Kapora, S., Hanssen, M., Niehof, J., Sandifort, Q.: Methodology for interference analysis during early design stages of high-performance mixed-signal ICs. In: Proceedings of 2015 10th International Workshop on the Electromagnetic Compatibility of Integrated Circuits (EMC Compo), Edinburgh, November 10–13, pp. 67–71 (2015)
9. Momentum Keysight Technologies. http://www.keysight.com/en/
10. Niehof, J., van Sinderen, J.: Preventing RFIC interference issues: a modeling methodology for floorplan development and verification of isolation- and grounding strategies. In: Proceedings of 2011 15th IEEE Workshop on Signal Propagation on Interconnects (SPI), Naples, pp. 11–14 (2011)

11. Putek, P., Crevecoeur, G., Slodička, M., van Keer, R., Van de Wiele, B., Dupré, L.: Space mapping methodology for defect recognition in eddy current testing—type NDT. COMPEL **31**(3), 881–894 (2012)
12. Putek, P., Meuris, P., Günther, M., ter Maten, E.J.W., Pulch, R., Wieers, A., Schoenmaker, W.: Uncertainty quantification in electro-thermal coupled problems based on a power transistor device. IFAC-PapersOnLine **48**, 938–939 (2015)
13. Putek, P., Janssen, R., Niehof, J., ter Maten, E.J.W., Pulch, R., Tasić, B., Günther, M.: Nanoelectronic coupled problem solutions: uncertainty quantification of RFIC interference. CASA-report, vol. 1622. Technische Universiteit Eindhoven, Eindhoven (2016)
14. Putek, P., Meuris, P., Pulch, R., ter Maten, E.J.W., Schoenmaker, W., Günther, M.: Uncertainty quantification for robust topology optimization of power transistor devices. IEEE Trans. Magn. **52**, 1700104 (2016)
15. Stanisavljevicć, M., Schmid, A., Leblebici, Y.: Reliability of Nanoscale Sircuits and Systems: Methodologies and Circuit Architectures. Springer, New York (2011)
16. Sudret, B.: Global sensitivity analysis using polynomial chaos expansion. Reliab. Eng. Syst. Saf. **93**(7), 964–979 (2008)
17. Xiu, D.: Numerical Methods for Stochastic Computations—A Spectral Method Approach. Princeton University Press, Princeton (2010)
18. Yildiz, Ö.F.: Analysis of electromagnetic interference variability on RF integrated circuits. MSc. thesis, Technische Universität Hamburg (2016)

Part V
Model Order Reduction

Model order reduction methods seek to reduce the complexity of a mathematical model in the process of its numerical solution. In this regard, finding and understanding the balance between complexity and accuracy certainly pose a challenging problem. Today, various methods for model order reduction have been developed. The keynote speakers Peter Benner and Lihong Feng, together with their co-authors, and the keynote speaker Roland Pulch made contributions to this topic in the form of two invited papers.

In the first invited paper on *"Sparse Model Order Reduction for Electro-Thermal Problems with Many Inputs"*, N. Banagaaya et al. propose a modified block-diagonal structured model order reduction method for electro-thermal coupled problems with multiple inputs. This modification yields sparse reduced-order models for both the electrical and thermal subsystems. Sparsity has a significant impact on the efficiency of the method.

The second invited paper, *"Quadrature Methods and Model Order Reduction for Sparse Approximations in Random Linear Dynamical Systems"* by R. Pulch, focuses on the construction of a sparse approximation to a quantity of interest derived from a linear dynamical system with random variables.

In turn, MD R. Hasan et al. investigate a *"POD-Based Reduced-Order Model of an Eddy-Current Levitation Problem"*. One approach is based on automatic remeshing. Further, the deformation of the finite element mesh of the subdomain around the moving body is considered. Both reduced-order approaches are compared with the full finite element simulation. Time-integration is always based on the implicit Euler method.

Lastly, R.V. Sabariego and J. Gyselinck discuss *"Time-Domain Reduced-Order Modeling of Linear Finite-Element Eddy-Current Problems via RL-Ladder Circuits"*, a reduced-order approach that they successfully apply to the TEAM 28 benchmark problem.

Chapter 18
Sparse Model Order Reduction for Electro-Thermal Problems with Many Inputs

Nicodemus Banagaaya, Lihong Feng, Wim Schoenmaker, Peter Meuris, Renaud Gillon, and Peter Benner

Abstract Recently, the block-diagonal structured model order reduction method for electro-thermal coupled problems with many inputs (BDSM-ET) was proposed in Banagaaya et al. (Model order reduction for nanoelectronics coupled problems with many inputs. In: Proceedings 2016 design, automation & test in Europe conference & exhibition, DATE 2016, Dresden, March 14–16, pp 313–318, 2016). After splitting the electro-thermal (ET) coupled problems into electrical and thermal subsystems, the BDSM-ET method reduces both subsystems separately, using Gaussian elimination and the block-diagonal structured MOR (BDSM) method, respectively. However, the reduced electrical subsystem has dense matrices and the nonlinear part of the reduced-order thermal subsystem is computationally expensive. We propose a modified BDSM-ET method which leads to sparser reduced-order models (ROMs) for both the electrical and thermal subsystems. Simulation of a very large-scale model with up to one million state variables shows that the proposed method achieves significant speed-up as compared with the BDSM-ET method.

N. Banagaaya (✉) · L. Feng · P. Benner
Max Planck Institute for Dynamics of Complex Technical Systems, Magdeburg, Germany
e-mail: banagaaya@mpi-magdeburg.mpg.de; feng@mpi-magdeburg.mpg.de;
benner@mpi-magdeburg.mpg.de

W. Schoenmaker · P. Meuris
Magwel NV, Leuven, Belgium
e-mail: wim.schoenmaker@magwel.com; peter.meuris@magwel.com

R. Gillon
ON Semiconductor Belgium, Oudenaarde, Belgium
e-mail: Renaud.Gillon@onsemi.com

© Springer International Publishing AG, part of Springer Nature 2018   189
U. Langer et al. (eds.), *Scientific Computing in Electrical Engineering*,
Mathematics in Industry 28, https://doi.org/10.1007/978-3-319-75538-0_18

18.1   Introduction

In several computational nanoelectronic problems, the spatial discretization of ET coupled problems leads to a nonlinear quadratic dynamical system of the following form:

$$\mathbf{E}\mathbf{x}'(t) = \mathbf{A}\mathbf{x}(t) + \mathbf{x}(t)^T \mathcal{F}\mathbf{x}(t) + \mathbf{B}\mathbf{u}(t), \quad \mathbf{x}(0) = \mathbf{x}_0, \tag{18.1a}$$

$$\mathbf{y}(t) = \mathbf{C}\mathbf{x}(t) + \mathbf{D}\mathbf{u}(t), \tag{18.1b}$$

where $\mathbf{E} \in \mathbb{R}^{n \times n}$ is singular, indicating that (18.1) is a system of differential-algebraic equations (DAEs), and $\mathbf{A} \in \mathbb{R}^{n \times n}$, $\mathbf{B} \in \mathbb{R}^{n \times m}$, $\mathbf{C} \in \mathbb{R}^{\ell \times n}$, $\mathbf{D} \in \mathbb{R}^{\ell \times m}$, while $\mathcal{F}$ is a 3-way tensor. n is called the order of the system, which is usually large. A tensor is a multi-way array and its order is the number of dimensions, also known as ways or modes, see [7]. Here, $\mathcal{F} = \left(\mathbf{F}_1^T, \ldots, \mathbf{F}_n^T\right)^T \in \mathbb{R}^{n \times n \times n}$ is a 3-way tensor of n matrices $\mathbf{F}_i \in \mathbb{R}^{n \times n}$. Each element in $\mathbf{x}(t)^T \mathcal{F}\mathbf{x}(t) \in \mathbb{R}^n$ is a scalar $\mathbf{x}(t)^T \mathbf{F}_i \mathbf{x}(t) \in \mathbb{R}$, $i = 1, \ldots, n$. The state vector $\mathbf{x}(t) = (\mathbf{x}_v(t)^T, \mathbf{x}_T(t)^T)^T \in \mathbb{R}^n$ includes the nodal voltages $\mathbf{x}_v(t) \in \mathbb{R}^{n_v}$, and the nodal temperatures $\mathbf{x}_T(t) \in \mathbb{R}^{n_T}$. $\mathbf{u}(t) \in \mathbb{R}^m$ and $\mathbf{y}(t) \in \mathbb{R}^\ell$ are the inputs (excitations) and the desired outputs (observations), respectively. We assume system (18.1) to be solvable, i.e., the matrix pencil $\lambda\mathbf{E} - \mathbf{A}$ is regular for all $\lambda \in \mathbb{C}$. In practice, more realistic models have very large dimension n compared to the number of inputs m and outputs ℓ. Despite the ever increasing computational power, simulation of these systems in acceptable time is still challenging. MOR aims to reduce the computational burden by generating ROMs that are faster and cheaper to simulate, yet accurately represent the original large-scale system behavior. MOR replaces (18.1) by a ROM

$$\mathbf{E}_r\mathbf{x}_r'(t) = \mathbf{A}_r\mathbf{x}_r(t) + \mathbf{x}_r(t)^T \mathcal{F}_r\mathbf{x}_r(t) + \mathbf{B}_r\mathbf{u}(t), \quad \mathbf{x}_r(0) = \mathbf{x}_{r0}, \tag{18.2a}$$

$$\mathbf{y_r}(t) = \mathbf{C}_r\mathbf{x}_r(t) + \mathbf{D}_r\mathbf{u}(t), \tag{18.2b}$$

where $\mathbf{E}_r, \mathbf{A}_r \in \mathbb{R}^{r \times r}$, $\mathbf{B}_r \in \mathbb{R}^{r \times m}$, $\mathbf{C}_r \in \mathbb{R}^{\ell \times r}$, $\mathbf{D}_r = \mathbf{D}$ and $\mathcal{F}_r \in \mathbb{R}^{r \times r \times r}$. $\mathbf{x}_r(t) \in \mathbb{R}^r$, $r \ll n$, is the reduced state vector and r is the order of the ROM. A good ROM should have small approximation error $\|\mathbf{y} - \mathbf{y}_r\|$ in a suitable norm $\| \cdot \|$ for every arbitrary input $\mathbf{u}(t)$. There exist many MOR methods for nonlinear (quadratic) systems such as the snapshot and implicit moment-matching methods, see [4] for a general discussion of MOR methods. The snapshot methods are not flexible for input-dependent systems as considered in this work, hence, we consider input-independent MOR methods, such as implicit moment-matching methods [4]. However, it is well known that as the number of inputs increases, the efficiency of moment-matching MOR methods decreases, since the size of the ROM is proportional to the number of inputs. Moreover, they cannot be applied directly to quadratic DAEs [3]. In general, models with numerous inputs and outputs are challenging for MOR, and most MOR methods produce large and dense ROMs for such systems. In [2], the BDSM-ET and SIP-ET methods for ET coupled problems with many inputs are proposed to overcome this problem. The BDSM-ET method is more accurate and leads to much smaller ROMs than the SIP-ET method. However, the BDSM-ET ROMs have dense matrices in the electrical subsystem and a dense 3-way tensor in the thermal subsystem, which restricts their applicability to small and medium sized ET systems. In this paper, we modify the BDSM-ET method proposed in [2]. In Sect. 18.2, we review the BDSM-ET method. Section 18.3 introduces the proposed modification of the BDSM-ET methods. Finally, we present numerical experiments and conclusions. For simplicity, we remove (t) for time dependent variables in the next sections.

18.2 BDSM-ET Method for ET Coupled Problems with Many Inputs

In this section, we discuss the BDSM-ET method proposed in [2]. We consider a structure arising naturally in nanoelectronic coupled problems with many inputs, taking the form of (18.1) with system matrices and tensor structures as below,

$$\mathbf{E} = \begin{pmatrix} 0 & 0 \\ 0 & \mathbf{E}_T \end{pmatrix}, \ \mathbf{A} = \begin{pmatrix} \mathbf{A}_v & 0 \\ 0 & \mathbf{A}_T \end{pmatrix}, \ \mathbf{B} = \begin{pmatrix} \mathbf{B}_v & 0 \\ 0 & \mathbf{B}_T \end{pmatrix}, \ \mathbf{C} = \begin{pmatrix} \mathbf{C}_v & \mathbf{C}_T \end{pmatrix}, \ \mathbf{D} = \begin{pmatrix} \mathbf{D}_v & \mathbf{D}_T \end{pmatrix},$$

$$\mathcal{F} = \left(0, \ldots, 0, \mathbf{F}_{n_v+1}^T, \ldots, \mathbf{F}_n^T\right)^T, \ \mathbf{F}_i = \begin{pmatrix} \mathbf{F}_{v_i} & 0 \\ 0 & 0 \end{pmatrix} \in \mathbb{R}^{n \times n}, i = n_v + 1, \ldots, n, \ \mathbf{u} = \begin{pmatrix} \mathbf{u}_v \\ \mathbf{u}_T \end{pmatrix},$$

with $\mathbf{A}_v \in \mathbb{R}^{n_v \times n_v}$, $\mathbf{B}_v \in \mathbb{R}^{n_v \times \tilde{m}}$, $\mathbf{E}_T \in \mathbb{R}^{n_T \times n_T}$, $\mathbf{A}_T \in \mathbb{R}^{n_T \times n_T}$, $\mathbf{B}_T \in \mathbb{R}^{n_T \times \tilde{m}}$, $\mathbf{C}_v \in \mathbb{R}^{\ell \times n_v}$, $\mathbf{F}_{v_i} \in \mathbb{R}^{n_v \times n_v}$ $\mathbf{C}_T \in \mathbb{R}^{\ell \times n_T}$, $\mathbf{D}_v \in \mathbb{R}^{\ell \times \tilde{m}}$, $\mathbf{D}_T \in \mathbb{R}^{\ell \times \tilde{m}}$, and $\mathbf{u}_v, \mathbf{u}_T \in \mathbb{R}^{\tilde{m}}$, $\tilde{m} = m/2$. Thus, substituting the above matrices and the tensor $\mathcal{F}$ into (18.1) leads to an equivalent decoupled system given by

$$\mathbf{A}_v \mathbf{x}_v = -\mathbf{B}_v \mathbf{u}_v, \tag{18.3a}$$

$$\mathbf{E}_T \mathbf{x}_T' = \mathbf{A}_T \mathbf{x}_T + \mathbf{x}_v^T \mathcal{F}_T \mathbf{x}_v + \mathbf{B}_T \mathbf{u}_T, \quad \mathbf{x}_T(0) = \mathbf{x}_{T_0}, \tag{18.3b}$$

$$\mathbf{y} = \mathbf{C}_v \mathbf{x}_v + \mathbf{C}_T \mathbf{x}_T + \mathbf{D}_v \mathbf{u}_v + \mathbf{D}_T \mathbf{u}_T, \tag{18.3c}$$

with $\mathcal{F}_T = \left(\mathbf{F}_{T_1}^T, \ldots, \mathbf{F}_{T_{n_T}}^T\right)^T \in \mathbb{R}^{n_v \times n_v \times n_T}$, $\mathbf{F}_{T_j} = \mathbf{F}_{v_{n_v+j}}, j = 1, \ldots, n_T$ and $\mathbf{F}_{v_i}$ is as defined earlier. Equations (18.3a) and (18.3b) are the electrical and thermal subsystems, respectively. After decoupling, the system (18.3) is now a one-way coupled system. Since the solution of the electrical and thermal subsystems can be computed consecutively, we call it decoupled, in contrast to the fully coupled original system, for which the electrical and the thermal subsystem must be solved simultaneously. We can observe that the nonlinear term $\mathbf{x}_v^T \mathcal{F}_T \mathbf{x}_v$ can be treated as part of the thermal input, since it is obtained by first simulating the electrical subsystem. The output can be obtained through (18.3c). Even after the above simplification, system (18.3) is still computationally expensive to simulate. Moreover, the decoupled system still has numerous inputs for both the electrical and the thermal subsystems. MOR replaces the decoupled system (18.3) with a reduced-order decoupled system

$$\mathbf{A}_{v_r}\mathbf{x}_{v_r} = -\mathbf{B}_{v_r}\mathbf{u}_v, \tag{18.4a}$$

$$\mathbf{E}_{T_r}\mathbf{x}'_{T_r} = \mathbf{A}_{T_r}\mathbf{x}_{T_r} + \mathbf{x}^T_{v_r}\mathcal{F}_{T_r}\mathbf{x}_{v_r} + \mathbf{B}_{T_r}\mathbf{u}_T, \quad \mathbf{x}_{T_r}(0) = \mathbf{x}_{T_{r_0}}, \tag{18.4b}$$

$$\mathbf{y}_r = \mathbf{C}_{v_r}\mathbf{x}_{v_r} + \mathbf{C}_{T_r}\mathbf{x}_{T_r} + \mathbf{D}_v\mathbf{u}_v + \mathbf{D}_T\mathbf{u}_T, \tag{18.4c}$$

where $\mathbf{A}_{v_r} \in \mathbb{R}^{r_v \times r_v}$, $\mathbf{B}_{v_r} \in \mathbb{R}^{r_v \times \tilde{m}}$, $\mathbf{E}_{T_r} \in \mathbb{R}^{r_T \times r_T}$, $\mathbf{A}_{T_r} \in \mathbb{R}^{r_T \times r_T}$, $\mathbf{B}_{T_r} \in \mathbb{R}^{r_T \times \tilde{m}}$, $\mathbf{C}_{v_r} \in \mathbb{R}^{\ell \times r_v}$, $\mathbf{C}_{T_r} \in \mathbb{R}^{\ell \times r_T}$, $\mathcal{F}_{T_r} \in \mathbb{R}^{r_v \times r_v \times r_T}$, with the reduced order $r = r_v + r_T \ll n$. In order to obtain the ROM (18.4), we combine the MOR techniques for algebraic and differential subsystems to obtain (18.4a) and (18.4b), respectively. MOR for general algebraic systems is still underdeveloped and the existing methods are often application specific, such as the method based on Gaussian elimination for algebraic systems arising from circuit simulations, see [5, 6, 9, 10] for details. MOR methods based on Gaussian elimination could be applied to algebraic systems, if the input matrix $\mathbf{B}_v$ has many zero rows, see [2]. The most challenging step is to reduce the nonlinear term in the thermal subsystem. The BDSM-ET method [2] was proposed to overcome this problem for the case of ET coupled problems which can be written in the form of (18.3). This method combines the Gaussian elimination based methods, such as SIP [10], with the BDSM method [11] to reduce the electrical and thermal subsystems, respectively. This can be briefly described as follows. Assume that $\mathbf{B}_v$ has many zero rows, then the electrical subsystem (18.3a) can be reformulated and partitioned as

$$\begin{pmatrix} \mathbf{A}_{v_{11}} & \mathbf{A}_{v_{12}} \\ \mathbf{A}^T_{v_{12}} & \mathbf{A}_{v_{22}} \end{pmatrix} \begin{pmatrix} \mathbf{x}_{v_e} \\ \mathbf{x}_{v_I} \end{pmatrix} = -\begin{pmatrix} \mathbf{B}_{v_e} \\ 0 \end{pmatrix} \mathbf{u}_v, \quad \mathbf{y}_v = \begin{pmatrix} \mathbf{C}_{v_e} & 0 \end{pmatrix} \begin{pmatrix} \mathbf{x}_{v_e} \\ \mathbf{x}_{v_I} \end{pmatrix} + \mathbf{D}_v\mathbf{u}_v, \tag{18.5}$$

where $\mathbf{x}_{v_e} \in \mathbb{R}^{n_{v_e}}$ and $\mathbf{x}_{v_I} \in \mathbb{R}^{n_{v_I}}$ represent the port and the internal nodal voltages, respectively, and $n_v = n_{v_e} + n_{v_I}$. Eliminating all internal nodes from (18.5) leads to the reduced-order electrical subsystem (18.4a) with matrix coefficients

$$\mathbf{A}_{v_r} = \left[\mathbf{A}_{v_{11}} - \mathbf{A}_{v_{12}}\mathbf{W}_v\right] \in \mathbb{R}^{r_v \times r_v}, \; \mathbf{B}_{v_r} = \mathbf{B}_{v_e} \in \mathbb{R}^{r_v \times \tilde{m}}, \; \mathbf{C}_{v_r} = \mathbf{C}_{v_e} \in \mathbb{R}^{\ell \times r_v}, \tag{18.6}$$

where $\mathbf{W}_v = \mathbf{A}^{-1}_{v_{22}}\mathbf{A}^T_{v_{12}} \in \mathbb{R}^{n_{v_I} \times n_{v_e}}$, $\mathbf{x}_{v_r} = \mathbf{x}_{v_e} \in \mathbb{R}^{r_v}$, and the order of the reduced electrical subsystem $r_v = n_{v_e} \ll n_v$. The reduction is based on the assumption that the input matrix $\mathbf{B}_v$ is very sparse in the sense that it has much fewer nonzero rows than the total row number, i.e. $n_{v_e} \ll n_v$. According to [11], the reduced matrix $\mathbf{A}_{v_r}$ is the Schur complement of the block $\mathbf{A}_{v_{22}}$ of the matrix $\mathbf{A}_v$. However, the Schur complement is dense due to the large number of fill-in. In many cases, eliminating all internal nodes at once is not advisable because it makes the construction of $\mathbf{W}_v = \mathbf{A}_{v_{22}}^{-1} \mathbf{A}_{v_{12}}^T$ responsible for the reduction, either costly or infeasible, since the matrix $\mathbf{A}_{v_{22}}$ can be very large due to a large number of internal nodes. It then produces a ROM (18.6) with very dense matrix $\mathbf{A}_{v_r}$ which may even be more computationally expensive than the original model. A sparse $\mathbf{A}_{v_r}$ can be obtained using sparsity control algorithms such as reduceR [9], which minimizes fill-in in the reduced matrix $\mathbf{A}_{v_r}$ by using fill-in reducing reordering algorithms, e.g., approximation minimum degree (AMD) [1], so that internal nodes responsible for fill-in are placed toward the end of the elimination sequence, along with the other nodes.

The reduction in the electrical subsystem induces a reduction in the thermal subsystem through the nonlinear part, leading to

$$\mathbf{E}_T \mathbf{x}_T' = \mathbf{A}_T \mathbf{x}_T + \mathbf{x}_{v_r}^T \tilde{\mathcal{F}}_T \mathbf{x}_{v_r} + \mathbf{B}_T \mathbf{u}_T, \quad \mathbf{x}_T(0) = \mathbf{x}_{T_0},$$

$$\mathbf{y}_T = \mathbf{C}_T \mathbf{x}_T + \mathbf{D}_T \mathbf{u}_T,$$

(18.7)

where $\tilde{\mathcal{F}}_T = \mathcal{F}_{T_{11}} - \mathbf{W}_v^T \mathcal{F}_{T_{21}} - \mathcal{F}_{T_{12}} \mathbf{W}_v + \mathbf{W}_v^T \mathcal{F}_{T_{22}} \mathbf{W}_v \in \mathbb{R}^{r_v \times r_v \times n_T}$ is a 3-way tensor. The 3-way tensors $\mathcal{F}_{T_{11}} \in \mathbb{R}^{n_{ve} \times n_{ve} \times n_T}$, $\mathcal{F}_{12} \in \mathbb{R}^{n_{ve} \times n_{vI} \times n_T}$, $\mathcal{F}_{21} \in \mathbb{R}^{n_{vI} \times n_{ve} \times n_T}$, $\mathcal{F}_{22} \in \mathbb{R}^{n_{vI} \times n_{vI} \times n_T}$ are the partitions of the tensor $\mathcal{F}_T$ corresponding to the partitions in (18.5). The next step is to apply the superposition principle to (18.7). Assume that the thermal input matrix $\mathbf{B}_T$ has no zero columns, so that it can be split into $\mathbf{B}_T = \sum_{i=1}^{\tilde{m}} \mathbf{B}_{T_i}$, where $\mathbf{B}_{T_i} \in \mathbb{R}^{n_T \times \tilde{m}}$ are column rank-1 matrices defined as

$$\mathbf{B}_{T_i}(:, j) = \begin{cases} \mathbf{b}_{T_i} \in \mathbb{R}^{n_T}, & \text{if } j = i, \\ 0, & \text{otherwise,} \end{cases} \quad i = 1, \ldots, \tilde{m}.$$

Here and below, blkdiag denotes the block-diagonal matrix defined by the input arguments. Applying the two-stage superposition principle from [2] to (18.7) leads to a block-diagonal structured system of dimension $\tilde{m} n_T$ given by

$$\mathcal{E}_T \tilde{\mathbf{x}}_T' = \mathcal{A}_T \tilde{\mathbf{x}}_T + \mathbf{x}_{v_r}^T \mathcal{F}_T \mathbf{x}_{v_r} + \mathcal{B}_T \mathbf{u}_T, \quad \tilde{\mathbf{x}}_T(0) = \left[\mathbf{x}_T(0), 0\right]^T,$$

$$\mathbf{y}_T = \mathcal{C}_T \tilde{\mathbf{x}}_T + \mathbf{D}_T \mathbf{u}_T,$$

(18.8)

where $\mathcal{E}_T = \text{blkdiag}(\mathbf{E}_T, \ldots, \mathbf{E}_T) \in \mathbb{R}^{\tilde{m} n_T \times \tilde{m} n_T}$, $\mathcal{C}_T = (\mathbf{C}_T, \ldots, \mathbf{C}_T) \in \mathbb{R}^{\ell \times \tilde{m} n_T}$, $\mathcal{A}_T = \text{blkdiag}(\mathbf{A}_T, \ldots, \mathbf{A}_T) \in \mathbb{R}^{\tilde{m} n_T \times \tilde{m} n_T}$, $\mathcal{B}_T = (\mathbf{B}_{T_1}^T, \ldots, \mathbf{B}_{T_{\tilde{m}}}^T)^T \in \mathbb{R}^{\tilde{m} n_T \times \tilde{m}}$, and $\mathcal{F}_T = \begin{pmatrix} \tilde{\mathcal{F}}_T \\ 0 \end{pmatrix} \in \mathbb{R}^{r_v \times r_v \times \tilde{m} n_T}$. The corresponding reduced-order thermal subsystem in the form of (18.4b) has block-diagonal structured matrices given by

$$\mathbf{E}_{T_r} = \mathbf{V}^T \mathcal{E}_T \mathbf{V}, \quad \mathbf{A}_{T_r} = \mathbf{V}^T \mathcal{A}_T \mathbf{V}, \quad \mathbf{B}_{T_r} = \mathbf{V}^T \mathcal{B}_T, \quad \mathbf{C}_{T_r} = \mathcal{C}_T \mathbf{V}, \quad (18.9)$$

where $\mathbf{V} = \mathrm{blkdiag}(\mathbf{V}^{(1)}, \ldots, \mathbf{V}^{(\tilde{m})})$. The projection matrices $\mathbf{V}^{(i)}$ can be constructed from each subsystem of (18.8) as (see [2] for details)

$$\mathrm{range}(\mathbf{V}^{(i)}) = \mathrm{span}\{\mathbf{R}_i, \mathbf{M}\mathbf{R}_i, \ldots, \mathbf{M}^{r_{T_i}-1}\mathbf{R}_i\}, \quad r_{T_i} \ll n_T, \qquad (18.10)$$

where $\mathbf{M} = (s_0\mathbf{E}_T - \mathbf{A}_T)^{-1}\mathbf{E}_T \in \mathbb{R}^{n_T \times n_T}$, and $\mathbf{R}_i = (s_0\mathbf{E}_T - \mathbf{A}_T)^{-1}\mathbf{b}_{T_i} \in \mathbb{R}^{n_T}$, $i = 1, \ldots, \tilde{m}$. The nonlinear term $\mathbf{V}^T\left(\mathbf{x}_{v_r}^T \mathscr{F}_T \mathbf{x}_{v_r}\right)$ can be reformulated as a reduced-order nonlinear term $\mathbf{x}_{v_r}^T \mathscr{F}_{T_r} \mathbf{x}_{v_r}$ using the following proposition from [3].

Proposition 18.1 *Let* $\mathbf{W} = \left(\mathbf{w}_{ij}\right) \in \mathbb{R}^{n \times r}$ *be a matrix,* $\mathbf{x}_r \in \mathbb{R}^r$, *and* $\tilde{\mathscr{F}} = \left[\tilde{\mathbf{F}}_1^T, \ldots, \tilde{\mathbf{F}}_n^T\right]^T \in \mathbb{R}^{r \times r \times n}$ *be a 3D tensor, then there exist a 3D tensor* $\mathscr{F}_r \in \mathbb{R}^{r \times r \times r}$, *such that:*

$$\mathbf{W}^T\left(\mathbf{x}_r^T \tilde{\mathscr{F}} \mathbf{x}_r\right) = \mathbf{x}_r^T \mathscr{F}_r \mathbf{x}_r,$$

where $\mathscr{F}_r = \left[\mathbf{F}_{r_1}^T, \ldots, \mathbf{F}_{r_r}^T\right]^T$ *with* $\mathbf{F}_{r_j} = \displaystyle\sum_{i=1}^{n} \mathbf{w}_{ij}\tilde{\mathbf{F}}_i \in \mathbb{R}^{r \times r}$, $j = 1, \ldots, r$.

From Proposition 18.1, we see that $\mathscr{F}_r$ in the reduced-order nonlinear term is independent of the time t and can be precomputed before simulating the ROM. Therefore reformulating the nonlinear term further improves the efficiency of simulating the ROM. It can be seen that $\mathbf{V}^{(i)}$ depends only on the single column $\mathbf{b}_{T_i}$, rather than $\mathbf{B}_T$ with many columns, leading to a block-wise sparse ROM as compared with the standard moment-matching methods, such as PRIMA [8]. Here, $s_0 \in \mathbb{C}$ is chosen arbitrarily. Finally, the order of the reduced thermal subsystem (18.4b) is $r_T = \sum_{i=1}^{\tilde{m}} r_{T_i}$. From the analysis in [2, 11], the block-diagonal system (18.8) yields a system equivalent to (18.7), so that the block-diagonal ROM of (18.8) can be considered as the ROM of (18.7). However, the matrix $\mathbf{A}_{v_r}$ and the tensor $\mathbf{F}_{T_r}$ in the ROM are dense which is still a computational and storage burden. In the next section, we propose a modified BDSM-ET method which leads to sparser ROMs.

18.3   Proposed Modified BDSM-ET Method

In this section, we propose the modified BDSM-ET method. The goal of the modified BDSM-ET method is to reduce the computational and storage demand of simulating the reduced electrical subsystem and the reduced nonlinear term in the thermal subsystem, obtained using the BDSM-ET method. Actually, the BDSM method in [11] can be extended to the electrical subsystem in algebraic form. Assume that the electrical input matrix $\mathbf{B}_v$ has no zero columns, so that it can be split into $\mathbf{B}_v = \sum_{i=1}^{\tilde{m}} \mathbf{B}_{v_i}$, where $\mathbf{B}_{v_i} \in \mathbb{R}^{n_v \times \tilde{m}}$ is a column rank-1 matrix defined as

$$
\mathbf{B}_{v_i}(:, j) = \begin{cases} \mathbf{b}_{v_i} \in \mathbb{R}^{n_v}, & \text{if } j = i, \\ 0, & \text{otherwise,} \end{cases} \quad i = 1, \dots, \tilde{m}.
$$

Applying the superposition principle to the electrical subsystem in (18.3) results in an equivalent block-diagonal algebraic system

$$
\mathscr{A}_v \xi_v = -\mathscr{B}_v \mathbf{u}_v, \quad \mathbf{y}_v = \mathscr{C}_v \xi_v, \tag{18.11}
$$

where $\mathscr{A}_v = \text{blkdiag}(\mathbf{A}_v, \dots, \mathbf{A}_v)$, $\mathscr{B}_v = (\mathbf{B}_{v_1}^T, \dots, \mathbf{B}_{v_{\tilde{m}}}^T)^T$, $\mathscr{C}_v = (\mathbf{C}_v, \dots, \mathbf{C}_v)$, $\xi_v = (\mathbf{x}_{v_1}^T, \dots, \mathbf{x}_{v_{\tilde{m}}}^T)^T$. The next step is to reduce the dimension of (18.11). This is done by applying reordering and elimination techniques to each subsystem of (18.11):

$$
\mathbf{A}_v \mathbf{x}_{v_i} = -\mathbf{B}_{v_i} \mathbf{u}_v, \quad \mathbf{y}_{v_i} = \mathbf{C}_v \mathbf{x}_{v_i}, \ i = 1, \dots, \tilde{m}. \tag{18.12}
$$

Assuming each $\mathbf{B}_{v_i}$ has many zero rows, then each subsystem in (18.12) can be reformulated as

$$
\begin{pmatrix} \mathbf{A}_{v11}^{(i)} & \mathbf{A}_{v12}^{(i)} \\ \mathbf{A}_{v12}^{(i)^T} & \mathbf{A}_{v22}^{(i)} \end{pmatrix} \begin{pmatrix} \mathbf{x}_{v_e}^{(i)} \\ \mathbf{x}_{v_I}^{(i)} \end{pmatrix} = -\begin{pmatrix} \mathbf{B}_{v_e}^{(i)} \\ 0 \end{pmatrix} \mathbf{u}_v, \quad \mathbf{y}_{v_i} = \begin{pmatrix} \mathbf{C}_{v_e}^{(i)} & 0 \end{pmatrix} \begin{pmatrix} \mathbf{x}_{v_e}^{(i)} \\ \mathbf{x}_{v_I}^{(i)} \end{pmatrix}, \tag{18.13}
$$

where $\mathbf{x}_{v_e}^{(i)} \in \mathbb{R}^{n_{v_e}^{(i)}}$ and $\mathbf{x}_{v_I}^{(i)} \in \mathbb{R}^{n_{v_I}^{(i)}}$ represent the port and the internal nodal voltages, respectively, and $n_v = n_{v_e}^{(i)} + n_{v_I}^{(i)}$, $i = 1, \dots, \tilde{m}$. Eliminating all internal nodes from (18.13) leads to the ROM of each subsystem as below

$$
\mathbf{A}_{v_{r_i}} \mathbf{x}_{v_{r_i}} = \mathbf{B}_{v_{r_i}} \mathbf{u}_v, \quad \mathbf{y}_{v_{r_i}} = \mathbf{C}_{v_{r_i}} \mathbf{x}_{v_{r_i}}, \tag{18.14}
$$

where $\mathbf{A}_{v_{r_i}} = \begin{bmatrix} \mathbf{A}_{v11}^{(i)} - \mathbf{A}_{v12}^{(i)} \mathbf{W}_{v_i} \end{bmatrix} \in \mathbb{R}^{r_{v_i} \times r_{v_i}}$, $\mathbf{B}_{v_{r_i}} = -\mathbf{B}_{v_e}^{(i)} \in \mathbb{R}^{r_{v_i} \times \tilde{m}}$, $\mathbf{C}_{v_{r_i}} = \mathbf{C}_{v_e}^{(i)} \in \mathbb{R}^{\ell \times r_{v_i}}$, $\mathbf{W}_{v_i} = \mathbf{A}_{v22}^{(i)^{-1}} \mathbf{A}_{v12}^{(i)^T} \in \mathbb{R}^{n_{v_I}^{(i)} \times n_{v_e}^{(i)}}$, $\mathbf{x}_{v_{r_i}} = \mathbf{x}_{v_e}^{(i)} \in \mathbb{R}^{r_{v_i}}$, and $r_{v_i} = n_{v_e}^{(i)} \ll n_v$. Replacing each $\mathbf{A}_v, \mathbf{B}_{v_i}, \mathbf{C}_v, \mathbf{x}_{v_i}$ in (18.11) with $\mathbf{A}_{v_{r_i}}, \mathbf{B}_{v_{r_i}}, \mathbf{C}_{v_{r_i}}, \mathbf{x}_{v_{r_i}}$ leads to the ROM of (18.11), which is also the ROM of (18.3a) of dimension $r_v = \sum_{i=1}^{\tilde{m}} r_{v_i}$ and with matrices

$$
\mathbf{A}_{v_r} = \text{blkdiag}(\mathbf{A}_{v_{r_1}}, \dots, \mathbf{A}_{v_{r_{\tilde{m}}}}), \ \mathbf{B}_{v_r} = (\mathbf{B}_{v_{r_1}}^T, \dots, \mathbf{B}_{v_{r_{\tilde{m}}}}^T)^T, \ \mathbf{C}_{v_r} = (\mathbf{C}_{v_{r_1}}, \dots, \mathbf{C}_{v_{r_{\tilde{m}}}}).
$$

Finally, we reduce the thermal subsystem (18.3b). Here, we propose the approach which leads to a much sparser reduced 3-way tensor than that obtained using the BDSM-ET method. Applying the superposition principle to the algebraic subsystem

(18.3a) introduces $\left(\sum_{i=1}^{\tilde{m}} \mathbf{x}_{v_i}^T\right) \mathcal{F}_T \left(\sum_{i=1}^{\tilde{m}} \mathbf{x}_{v_i}\right)$ into the thermal subsystem, i.e. $\mathbf{x}_v$ is replaced by $\sum_{i=1}^{\tilde{m}} \mathbf{x}_{v_i}$ in the nonlinear part. In order to obtain a sparse tensor, the approximation $\left(\sum_{i=1}^{\tilde{m}} \mathbf{x}_{v_i}^T\right) \mathcal{F}_T \left(\sum_{i=1}^{\tilde{m}} \mathbf{x}_{v_i}\right) \approx \sum_{i=1}^{\tilde{m}} \mathbf{x}_{v_i}^T \mathcal{F}_T \mathbf{x}_{v_i}$ is introduced for the thermal subsystem. From numerical simulations results, we have observed that the error introduced by the approximation is very small and can be neglected for the nanoelectronic problems considered.

Thus (18.3b) can be approximated as

$$\mathbf{E}_T \mathbf{x}_T' = \mathbf{A}_T \mathbf{x}_T + \xi_v^T \mathcal{F}_T \xi_v, +\mathbf{B}_T \mathbf{u}_T, \quad \mathbf{x}_T(0) = \mathbf{x}_{T_0}, \tag{18.15a}$$

$$\mathbf{y}_T = \mathbf{C}_T \mathbf{x}_T + \mathbf{D}_T \mathbf{u}_T. \tag{18.15b}$$

Here we have used the equality

$$\sum_{i=1}^{\tilde{m}} \mathbf{x}_{v_i}^T \mathcal{F}_T \mathbf{x}_{v_i} = \xi_v^T \mathcal{F}_T \xi_v,$$

where $\mathcal{F}_T = \left[\mathcal{F}_{T_1}^T, \ldots, \mathcal{F}_{T_{n_T}}^T\right]^T \in \mathbb{R}^{\tilde{n}_v \times \tilde{n}_v \times n_T}$, $\tilde{n}_v = \tilde{m} n_v$, $\mathcal{F}_{T_i} = \text{blkdiag}(\mathbf{F}_{T_i}, \ldots, \mathbf{F}_{T_i}) \in \mathbb{R}^{\tilde{n}_v \times \tilde{n}_v}$, $\mathbf{F}_{T_i} \in \mathbb{R}^{n_v \times n_v}$ and ξ_v is defined as in (18.11). We can see that each reduced state in (18.14) induces a reduction in (18.15) leading to

$$\mathbf{E}_T \mathbf{x}_T' = \mathbf{A}_T \mathbf{x}_T + \xi_{v_r}^T \mathcal{F}_{T_r} \xi_{v_r} + \mathbf{B}_T \mathbf{u}_T, \quad \mathbf{x}_T(0) = \mathbf{x}_{T_0}, \tag{18.16a}$$

$$\mathbf{y}_T = \mathbf{C}_T \mathbf{x}_T + \mathbf{D}_T \mathbf{u}_T, \tag{18.16b}$$

where $\xi_{v_r} = (\mathbf{x}_{v_{r_1}}^T, \ldots, \mathbf{x}_{v_{r_{\tilde{m}}}}^T)^T$, $\mathcal{F}_{T_r} = \left[\mathcal{F}_{T_{r_1}}^T, \ldots, \mathcal{F}_{T_{r_{n_T}}}^T\right]^T \in \mathbb{R}^{r_v \times r_v \times n_T}$, with $\mathcal{F}_{T_{r_i}} = \text{blkdiag}(\mathbf{F}_{T_{r_i}}, \ldots, \mathbf{F}_{T_{r_i}}) \in \mathbb{R}^{r_v \times r_v}$, where $\mathbf{F}_{T_{r_i}} = \mathbf{F}_{T_{11}}^{(i)} - \mathbf{W}_{v_i}^T \mathbf{F}_{T_{21}}^{(i)} - \mathbf{F}_{T_{12}}^{(i)} \mathbf{W}_{v_i} + \mathbf{W}_{v_i}^T \mathbf{F}_{T_{22}}^{(i)} \mathbf{W}_{v_i} \in \mathbb{R}^{r_{v_i} \times r_{v_i}}$. Here $\mathbf{F}_{T_{11}}^{(i)}, \mathbf{F}_{T_{12}}^{(i)}, \mathbf{F}_{T_{21}}^{(i)}, \mathbf{F}_{T_{22}}^{(i)}$ are the sub-blocks of $\mathbf{F}_{T_i}$ partitioned according to the partition of $\mathbf{A}_v$ in (18.13). Since $\sum_{i=1}^{\tilde{m}} \mathbf{x}_{v_i}^T \mathcal{F}_T \mathbf{x}_{v_i}$ can be considered as an extra input for the thermal subsystem, the superposition principle still applies to the thermal subsystem. Therefore, (18.16) can also be split into $\tilde{m}$ subsystems, the thermal state $\mathbf{x}_T$ of (18.16) can be reduced following the steps from (18.8) till the end of Sect. 18.2. The reduced thermal system is in the form of (18.4b) with the reduced matrices being defined in (18.9). Using Proposition 18.1, the nonlinear term $\mathbf{V}^T \left(\xi_{v_r}^T \tilde{\mathcal{F}}_T \xi_{v_r}\right)$, where $\tilde{\mathcal{F}}_T = \begin{pmatrix} \mathcal{F}_{T_r} \\ 0 \end{pmatrix} \in \mathbb{R}^{r_v \times r_v \times \tilde{m} n_T}$, $\tilde{\mathcal{F}}_T = \left[\tilde{\mathcal{F}}_{T_1}^T, \ldots, \tilde{\mathcal{F}}_{T_{\tilde{m} n_T}}^T\right]^T$ with $\tilde{\mathcal{F}}_{T_i} \in \mathbb{R}^{r_v \times r_v}$ can also be reformulated as $\xi_{v_r}^T \tilde{\mathcal{F}}_{T_r} \xi_{v_r}$, where $\tilde{\mathcal{F}}_{T_r} = \left[\tilde{\mathcal{F}}_{T_{r_1}}^T, \ldots, \tilde{\mathcal{F}}_{T_{r_{r_T}}}^T\right]^T \in \mathbb{R}^{r_v \times r_v \times r_T}$ with $\tilde{\mathcal{F}}_{T_{r_j}} = \sum_{i=1}^{\tilde{m} n_T} v_{ji} \tilde{\mathcal{F}}_{T_i} \in$

$\mathbb{R}^{r_v \times r_v}$, $j = 1, \ldots, r_T$, $\mathbf{V} = (\mathbf{v}_{ij}) \in \mathbb{R}^{\tilde{m} n_T \times r_T}$. Here, the reduction matrix $\mathbf{V}$ is defined and computed as in (18.10). Instead of a dense tensor as in the previous section, here $\tilde{\mathscr{F}}_{T_r}$ is in block-diagonal form which is sparse. Combining the above block structured reduced electrical and thermal subsystems, we obtain the modified BDSM-ET ROMs of (18.1) in the form of (18.2) with system matrices

$$
\mathbf{E}_r = \begin{pmatrix} 0 & 0 \\ 0 & \mathbf{E}_{T_r} \end{pmatrix}, \ \mathbf{A}_r = \begin{pmatrix} \mathbf{A}_{v_r} & 0 \\ 0 & \mathbf{A}_{T_r} \end{pmatrix}, \ \mathbf{B}_r = \begin{pmatrix} \mathbf{B}_{v_r} & 0 \\ 0 & \mathbf{B}_{T_r} \end{pmatrix}, \ \mathbf{C}_r = \begin{pmatrix} \mathbf{C}_{v_r} & \mathbf{C}_{T_r} \end{pmatrix}, \ \mathbf{D} = \begin{pmatrix} \mathbf{D}_v & \mathbf{D}_T \end{pmatrix},
$$

$$
\mathscr{F}_r = \left(0, \ldots, 0, \mathbf{F}_{r_v+1}^T, \ldots, \mathbf{F}_{r_v+r_T}^T\right)^T, \ \mathbf{F}_{r_v+j} = \begin{pmatrix} \tilde{\mathscr{F}}_{T_{r_j}} & 0 \\ 0 & 0 \end{pmatrix} \in \mathbb{R}^{r \times r}, \ j = 1, \ldots, r_T.
$$

Hence, by construction, the modified BDSM-ET method constructs sparser ROMs than the BDSM-ET method proposed in [2], since all its reduced matrices and the tensor are block-wise sparse as also illustrated in the next section.

18.4 Numerical Experiments

In this section, we illustrate the efficiency of the modified BDSM-ET method by examining three ET coupled models from industrial applications, namely, a package model ($n = 9193, m = 34, \ell = 68$), a power-MOS model ($n = 13,216, m = 6, \ell = 12$), and a power cell model ($n = 925,286, m = 408, \ell = 816$) as shown in Table 18.1. The first two ET models are nonlinear quadratic DAEs of the form (18.1), while the last model is a linear DAE, i.e., $\mathscr{F} = 0$. Simulations on the first two ET models are done in MATLAB®Version 2012b on a Laptop with 6 GB RAM, CPU@ 2.00 GHz. Simulation on the power cell model is done on a Unix compute server with 1 TB main memory.

All these models can be reformulated into an equivalent decoupled system of the form (18.3). Then, the numerical solutions are obtained by applying the built-in MATLAB function mldivide(/) to the electrical subsystem and the implicit-Euler integration scheme to the thermal subsystem in the desired time interval. We reduce each ET decoupled model using the PRIMA-ET, BDSM-ET and the proposed modified BDSM-ET methods. The PRIMA-ET method uses the Gaussian elimination and PRIMA methods, to reduce the order of the electrical and thermal subsystems, without applying the superposition principle. The other two MOR methods are as discussed in Sects. 18.2 and 18.3, respectively.

In Table 18.1, n_T is the order of the thermal subsystem, n_v is the order of the electrical subsystem, r_v is the order of the reduced electrical subsystem, r_T is the order of the reduced thermal subsystem, $r = r_v + r_T$ is the order of the reduced ET coupled model, "%Red" means the reduction rate in % w.r.t. the original order n. In Table 18.2, "Stor. (Mb)" is the storage requirement, "Error" is the maximum output relative error in time domain, "Speed-up" represents the speed-up factor w.r.t. the time for simulating the original large model. From Table 18.1, we can see that PRIMA-ET was unable to reduce the large model with dimension

Table 18.1 Dimension comparison of ROMs, $r = r_v + r_T$

Models			Decoupled models		PRIMA-ET				BDSM-ET				Modified BDSM-ET			
n	m	ℓ	n_v	n_T	r_v	r_T	r	% Red	r_v	r_T	r	% Red	r_v	r_T	r	% Red
9193	34	68	8071	1122	188	198	386	95.8	188	198	386	95.8	238	198	436	95.3
13,216	6	12	11,556	1660	160	63	223	98.3	160	84	244	98.15	160	84	244	98.15
925,286	408	816	392,773	532,513	–	–	–	–	9264	4305	13,569	98.53	9396	4305	13,701	98.52

Table 18.2 Efficiency comparison of ROMs

Models	PRIMA-ET				BDSM-ET				Modified BDSM-ET			
n	r	Stor. (Mb)	Error	Speedup	r	Stor. (Mb)	Error	Speed-up	r	Stor. (Mb)	Error	Speed-up
9193	386	140.3	7.2×10^{-9}	12.3	386	16	2.1×10^{-2}	65.7	436	2.6	2.1×10^{-2}	70.6
13,216	223	27.01	3.5×10^{-5}	74.3	244	27.03	1.4×10^{-2}	120	244	14.2	1.4×10^{-2}	157.1
925,286	–	–	–	–	13,569	385.3	6.3×10^{-8}	5.7	13,701	56.4	7.0×10^{-7}	972.7

925,286, because of memory limitations. Comparing the BDSM-ET type methods with the PRIMA-ET method, we see that both methods produce accurate ROMs with large speed-ups as shown in Table 18.2. The modified BDSM-ET ROMs are computationally cheaper than the BDSM-ET ROMs yet with almost the same accuracy, especially for large models. For the case of the power cell model, the modified BDSM-ET ROM is 170.6 faster than the BDSM-ET method. This is due to the fact that the resulting reduced model is completely block-wise sparse (see Fig. 18.4), and each block is very small w.r.t. the original order n, which results in a very sparse ROM. Furthermore, it requires much less storage requirements, since it constructs sparse ROMs as illustrated in Figs. 18.1, 18.2, 18.3 and 18.4. In Table 18.3, we compare the off-line costs which are the times to construct the ROMs. We can observe that modified BDSM-ET ROMs are computationally more expensive to construct compared to the other ROMs and their computational cost depends on the number of inputs.

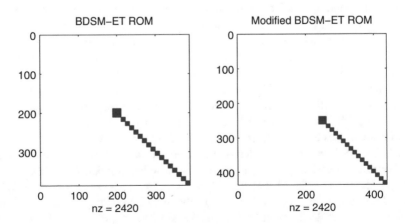

Fig. 18.1 Comparison of the sparsity of the reduced matrix $\mathbf{E}_r$, $n = 9193$

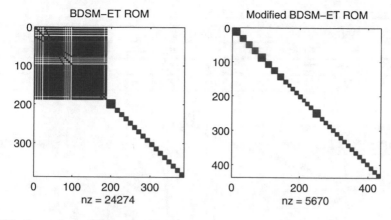

Fig. 18.2 Comparison of the sparsity of the reduced matrix $\mathbf{A}_r$, $n = 9193$

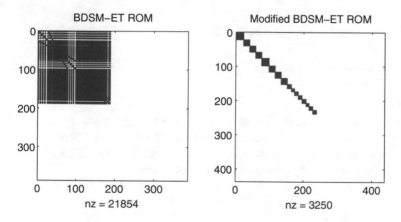

Fig. 18.3 Comparison of the sparsity of the first nonzero slice of the reduced tensor $\mathcal{F}_r$, $n = 9193$

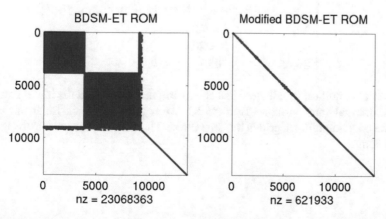

Fig. 18.4 Comparison of the sparsity of the reduced power cell matrix $\mathbf{A}_r$, $n = 925{,}286$

Table 18.3 Off-line cost comparison of ROMs

Models	PRIMA-ET		BDSM-ET		Modified BDSM-ET	
n	r	Off-line cost (s)	r	Offline-cost	r	Offline-cost (s)
9193	386	22	386	18.5	436	256.2
13,216	223	12.1	244	13	244	1172.7

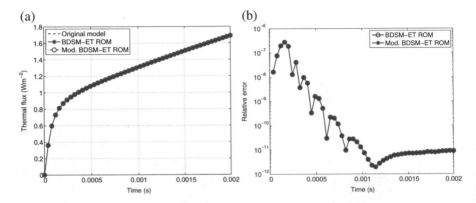

Fig. 18.5 Comparison of the outputs at port 611, $(\mathbf{y}_{611}, n = 925,286)$. (**a**) The thermal flux. (**b**) The relative error

In Fig. 18.5, we compare the outputs at port 611, $\mathbf{y}_{611}$, given by the BDSM-ET type ROMs and the original power-cell model. The power-cell model corresponds to a power-transistor design of ONN that is intended for use in smart-power ICs. The system is excited by 408 inputs defined as below.

$$
u_i = \begin{cases}
5, & 1 \leq i \leq 200, \\
0, & i = 201, \\
0, & i = 202, t \in [0, 10^{-7}), \\
1.5(10^7 t - 1), & i = 202, t \in [10^{-7}, 2 \times 10^{-7}], \\
1.5, & i = 202, t \in (2 \times 10^{-7}, 5 \times 10^{-7}], \\
10, & i = 203, \\
0, & i = 204, \\
26.85 & 205 \leq i \leq 408.
\end{cases}
$$

The initial condition for all electrical state variables is 0 V, and the initial condition for all thermal state variables is 26.85 °C. We used the implicit-Euler integration scheme on a nonuniform grid in the time interval [0, 0.002 s] to simulate the thermal subsystem.

Both methods introduce very small relative errors as shown in Fig. 18.5b. The ROM error is defined as

$$\max_{i \in \{t_1,\ldots,t_{29}\}} \| y_i - y_{r_i} \|_2 / \| y_i \|,$$

where $y_i \in \mathbb{R}^{n \times \ell}$ is the output, obtained from the original power-cell model, it is a vector containing all the output values at the ith nonuniform time step $t_i, i = 1, \ldots, 29$ in the time interval $[0, 2.0 \times 10^{-3} \, \text{s}]$.

18.5 Conclusion

We have proposed a modified BDSM-ET method for ET coupled problems with many inputs arising from industrial applications. The modified BDSM-ET method produces sparse yet accurate ROMs compared with the BDSM-ET method. Finally, the proposed method allows independent calculations which attracts parallelization. This could be a topic in the future.

Acknowledgements This work is supported by the collaborative project nanoCOPS, Nanoelectronics COupled Problems Solutions, supported by the European Union in the FP7-ICT-2013-11 Program under Grant Agreement Number 619166.

References

1. Amestoy, P., Davis, T., Duff, I.: An approximate minimum degree ordering algorithm. SIAM J. Matrix Anal. Appl. **17**(4), 886–905 (1996)
2. Banagaaya, N., Feng, L., Schoenmaker, W., Meuris, P., Wieers, A., Gillon, R., Benner, P.: Model order reduction for nanoelectronics coupled problems with many inputs. In: Proceedings 2016 Design, Automation & Test in Europe Conference & Exhibition, DATE 2016, Dresden, March 14–16, pp. 313–318 (2016)
3. Banagaaya, N., Feng, L., Schoenmaker, W., Meuris, P., Gillon, R., Benner, P.: An index-aware parametric model order reduction for parametrized quadratic DAEs. Appl. Math. Comput. (2018, in press). https://doi.org/10.1016/j.amc.2017.04.024
4. Baur, U., Benner, P., Feng, L.: Model order reduction for linear and nonlinear systems: a system-theoretic perspective. Arch. Comput. Meth. Eng. **21**(4), 331–358 (2014)
5. Ionutiu, R., Rommes, J., Schilders, W.: SparseRC: Sparsity preserving model reduction for RC circuits with many terminals. IEEE Trans. Comput. Aided Des. Integr. Circuits Syst. **30**(12), 1828–1841 (2011)
6. Kerns, K.J., Yang, A.T.: Stable and efficient reduction of large, multiport RC networks by pole analysis via congruence transformations. IEEE Trans. Comput. Aided Des. Integr. Circuits Syst. **16**(7), 734–744 (1997)
7. Kolda, T.G., Bader, B.W.: Tensor decompositions and applications. SIAM Rev. **51**(3), 455–500 (2009)

8. Odabasioglu, A., Celik, M., Pileggi, L.T.: PRIMA: passive reduced-order interconnect macro-modeling algorithm. IEEE Trans. Comput. Aided Des. Integr. Circuits Syst. **17**(8), 645–654 (1998)
9. Rommes, J., Schilders, W.: Efficient methods for large resistor networks. IEEE Trans. Comput. Aided Des. Integr. Circuits Syst. **29**(1), 28–39 (2010)
10. Ye, Z., Vasilyev, D., Zhu, Z., Phillips, J.: Sparse Implicit Projection (SIP) for reduction of general many-terminal networks. In Proceedings of International Conference on Computer-Aided Design (2008)
11. Zhang, Z., Hu, C.C.X., Wong, N.: A block-diagonal structured model reduction scheme for power grid networks. In Design, Automation & Test in Europe Conference & Exhibition (DATE) (2011)

Chapter 19
Quadrature Methods and Model Order Reduction for Sparse Approximations in Random Linear Dynamical Systems

Roland Pulch

Abstract We consider linear dynamical systems including random variables to model uncertainties of physical parameters. The output of the system is expanded into a series with orthogonal basis functions. Our aim is to identify a sparse approximation, where just a low number of basis functions is required for a sufficiently accurate representation. The coefficient functions of the expansion are approximated by a quadrature method or a sampling technique. The performance of a quadrature scheme can be described by a larger linear dynamical system, which is weakly coupled. We apply methods of model order reduction to the coupled system, which results in a sparse approximation of the original expansion. The approximation error is estimated by Hardy norms of transfer functions. Furthermore, we present numerical results for a test example modelling the electric circuit of a band pass filter.

19.1 Introduction

The mathematical modelling of electronic circuits and devices yields dynamical systems, which contain physical or geometrical parameters. The parameters often exhibit uncertainties or fuzziness. In nanoelectronics, for example, miniaturisation causes imperfections in an industrial production. An uncertainty quantification is often performed by a stochastic modelling, where uncertain parameters are replaced by random variables or random processes, see [24].

We discuss linear time-invariant dynamical systems with random variables. An output of the system is defined as a quantity of interest. The quantity of interest is expanded into a series with orthogonal basis functions. The emphasis is on polynomial bases due to the concept of the polynomial chaos, see [2, 24].

R. Pulch (✉)

Institute for Mathematics and Computer Science, Ernst-Moritz-Arndt-Universität Greifswald, Greifswald, Germany

e-mail: roland.pulch@uni-greifswald.de; pulchr@uni-greifswald.de

© Springer International Publishing AG, part of Springer Nature 2018   203
U. Langer et al. (eds.), *Scientific Computing in Electrical Engineering*,
Mathematics in Industry 28, https://doi.org/10.1007/978-3-319-75538-0_19

We want to identify a sparse approximation, where a sum with just a few basis functions is sufficiently accurate. For such low-dimensional approximations, least angle regression [5], sparse grid quadrature [8], compressed sensing [10] and ℓ_1-minimisation [14] were applied as a tool in previous works.

The unknown coefficient functions of the expansion can be approximated by either stochastic Galerkin methods or stochastic collocation schemes, see [18]. In the stochastic Galerkin method, a larger coupled linear dynamical system appears, which has to be solved once. Model order reduction (MOR) was applied successfully to this coupled system in [21, 22]. Details on MOR methods for linear dynamical systems can be found in [1, 4, 11], for example. Moreover, methods of MOR allow for the identification of a sparse representation by a reduction of the Galerkin system as shown in [19].

In this work, we apply stochastic collocation methods, which are defined by quadrature techniques or sampling schemes, to determine such a low-dimensional approximation. We consider the case of a relatively large number of random variables. Sparse grid constructions [12], cubature techniques [23] and quasi Monte-Carlo methods [17] are feasible. The original dynamical system has to be solved many times separately for each node of the quadrature method. However, the quadrature approach can also be formulated as a larger weakly coupled linear dynamical system, see [20]. Now we use MOR to this large auxiliary system to obtain a sparse approximation for the quantity of interest. The same error estimates are valid as in [19], which are derived by the Hardy norms of transfer functions. We analyse a new aspect of quadrature methods in the case of symmetric probability distributions.

The paper is organised as follows. The stochastic modelling, the orthogonal expansions and the problem of sparse approximations are outlined in Sect. 19.2. Numerical methods based on quadrature techniques are discussed in Sect. 19.3. MOR determines sparse approximations in Sect. 19.4. Finally, Sect. 19.5 shows numerical results for a test example, where a sparse grid quadrature and a quasi Monte-Carlo method are compared.

19.2 Stochastic Modelling and Sparse Representations

The problem under investigation is defined in this section.

19.2.1 Linear Dynamical Systems

We consider linear time-invariant dynamical systems

$$
\begin{aligned}
\mathbf{E}(\mathbf{p})\dot{\mathbf{x}}(t, \mathbf{p}) &= \mathbf{A}(\mathbf{p})\mathbf{x}(t, \mathbf{p}) + \mathbf{B}(\mathbf{p})u(t) \\
y(t, \mathbf{p}) &= \mathbf{C}(\mathbf{p})\mathbf{x}(t, \mathbf{p})
\end{aligned}
\tag{19.1}
$$

for $t \geq 0$ with the state variables or inner variables $\mathbf{x} \in \mathbb{R}^n$, the input signal $u \in \mathbb{R}$, the output variable $y \in \mathbb{R}$, the matrices $\mathbf{A}, \mathbf{E} \in \mathbb{R}^{n \times n}$, the column vector $\mathbf{B} \in \mathbb{R}^n$ and the row vector $\mathbf{C} \in \mathbb{R}^n$. Matrices and vectors include physical parameters $\mathbf{p} \in \Pi \subseteq \mathbb{R}^q$. Hence the state variables or inner variables as well as the output depend on both the time and the parameters. We discuss a single-input-single-output (SISO) system (19.1), while generalisations to multiple-input-multiple-output (MIMO) systems are straightforward. Initial values $\mathbf{x}(0, \mathbf{p}) = \mathbf{x}_0(\mathbf{p})$ are predetermined at $t = 0$.

On the one hand, the system (19.1) consists of ordinary differential equations (ODEs) in the case of a regular matrix $\mathbf{E}(\mathbf{p})$. On the other hand, a descriptor system of differential algebraic equations (DAEs) occurs for a singular matrix $\mathbf{E}(\mathbf{p})$. The mathematical modelling of linear electric circuits by modified nodal analysis [13], for example, typically generates singular matrices $\mathbf{E}(\mathbf{p})$ for all $\mathbf{p}$. We assume that the systems (19.1) are asymptotically stable, i.e., the eigenvalues of the matrix pencil $\lambda \mathbf{E}(\mathbf{p}) - \mathbf{A}(\mathbf{p})$ have negative real parts for each $\mathbf{p} \in \Pi$.

19.2.2 Stochastic Modelling and Orthogonal Expansions

The physical parameters often include uncertainties. A common approach consists in the substitution of the parameters by random variables, see [24]. Let the probability space $(\Omega, \mathscr{A}, \mu)$ be given with event space Ω, sigma-algebra $\mathscr{A}$ and probability measure μ. We assume that the random variables $\mathbf{p} : \Omega \to \Pi$ are independent. Furthermore, let a probability density function $\rho : \Pi \to \mathbb{R}$ be available. For a measurable function $f : \Pi \to \mathbb{R}$, the expected value reads as

$$\mathbb{E}[f] = \int_{\Omega} f(\mathbf{p}(\omega)) \, d\mu(\omega) = \int_{\Pi} f(\mathbf{p}) \rho(\mathbf{p}) \, d\mathbf{p} \qquad (19.2)$$

provided that the integral exists. The Hilbert space

$$\mathscr{L}^2(\Pi, \rho) := \left\{ f : \Pi \to \mathbb{R} \ : \ f \text{ measurable and } \mathbb{E}[f^2] < \infty \right\}$$

is equipped with the inner product $< f, g >:= \mathbb{E}[fg]$ using the expected value (19.2).

We assume that a complete orthonormal basis $(\Phi_i)_{i \in \mathbb{N}}$ is given in $\mathscr{L}^2(\Pi, \rho)$. It holds that $< \Phi_i, \Phi_j >= \delta_{ij}$ with the Kronecker delta. If $y(t, \cdot) \in \mathscr{L}^2(\Pi, \rho)$ for each $t \geq 0$, then the output can be expanded into the series

$$y(t, \mathbf{p}) = \sum_{i=1}^{\infty} w_i(t) \Phi_i(\mathbf{p}). \qquad (19.3)$$

The involved coefficient functions are defined by

$$w_i(t) := \ <y(t, \cdot), \Phi_i(\cdot)> \ = \int_\Pi y(t, \mathbf{p})\Phi_i(\mathbf{p})\rho(\mathbf{p}) \, d\mathbf{p}. \tag{19.4}$$

Often orthonormal polynomials are chosen associated to the theory of the polynomial chaos, see [2, 24]. A truncation of the series (19.3) results in

$$y^{(\mathbb{I})}(t, \mathbf{p}) := \sum_{i\in\mathbb{I}} w_i(t)\Phi_i(\mathbf{p}) \tag{19.5}$$

with a finite subset $\mathbb{I} \subset \mathbb{N}$. Numerical methods yield approximations

$$\hat{y}^{(\mathbb{I})}(t, \mathbf{p}) := \sum_{i\in\mathbb{I}} \hat{w}_i(t)\Phi_i(\mathbf{p}). \tag{19.6}$$

The involved coefficients functions $\hat{w}_i$ for $i \in \mathbb{I}$ can be computed by either stochastic Galerkin methods or stochastic collocation techniques, see [18, 24].

19.2.3 Sparse Approximations

In the case of a polynomial basis, the approximations (19.5), (19.6) often include all multivariate polynomials up to a total degree d, cf. [5, Sect. 2], which yields the index set

$$\mathbb{I}_d := \left\{ i \ : \ \Phi_i(p) = \phi_{j_1}^{(1)}(p_1) \cdots \phi_{j_q}^{(q)}(p_q) \ \text{with} \ j_1 + \cdots + j_q \le d \right\}. \tag{19.7}$$

The sequence $(\phi_j^{(\ell)})_{j\in\mathbb{N}_0}$ includes the univariate orthonormal polynomials with respect to the distribution of the ℓth random variable and the degree of $\phi_j^{(\ell)}$ is exactly j. The number of basis polynomials becomes, see [24, Eq. (5.24)],

$$|\mathbb{I}_d| = \frac{(q+d)!}{q!d!}. \tag{19.8}$$

This number is huge for large numbers q of parameters even if the total degree is moderate like $3 \le d \le 5$.

Given a large orthonormal basis $\{\Phi_1, \ldots, \Phi_m\}$ with $m := |\mathbb{I}|$ for some index set $\mathbb{I}$ as a starting point, our aim is to identify a small orthonormal basis $\{\Psi_1, \ldots, \Psi_r\}$ for some $r \ll m$ with an approximation

$$\bar{y}^{(r)}(t, \mathbf{p}) := \sum_{i=1}^r \bar{v}_i(t)\Psi_i(\mathbf{p}). \tag{19.9}$$

The total approximation error can be estimated by

$$
\left\| y - \overline{y}^{(r)} \right\|_{L^2(\Pi,\rho)} \leq \left\| y - y^{(\mathbb{I})} \right\|_{L^2(\Pi,\rho)} + \left\| y^{(\mathbb{I})} - \hat{y}^{(\mathbb{I})} \right\|_{L^2(\Pi,\rho)} + \left\| \hat{y}^{(\mathbb{I})} - \overline{y}^{(r)} \right\|_{L^2(\Pi,\rho)}
$$
$$(19.10)$$

for each time point separately using Eqs. (19.3), (19.5), (19.6) and (19.9). The right-hand side of (19.10) consists of the truncation error, the error of the numerical method and an additional sparsification error. Our focus is on the sparsification error. We assume that the first and second error term are sufficiently small by including enough basis functions and choosing a sufficiently accurate numerical method. In an optimal case, all three error terms should have the same order of magnitude.

For given $\varepsilon > 0$, let $d(\varepsilon)$ be the smallest integer such that $\mathbb{I}_{d(\varepsilon)}$ from (19.7) yields a truncation error below ε. We define the sparsity of the representation (19.9) similar to [5, Sect. 4.1] by the ratio $\sigma := r/|\mathbb{I}_{d(\varepsilon)}|$. If $\sigma \ll 1$ can be achieved with a sufficiently small approximation error, then the representation is efficient. In our context, the sparsity σ is time-dependent given some initial value problem of the system (19.1).

19.3 Quadrature Methods

We examine quadrature methods and sampling techniques for the approximation of unknown coefficient functions now.

19.3.1 Weakly Coupled Linear Dynamical System

The task consists in the computation of the unknown coefficient functions (19.4) for without loss of generality $i = 1, \ldots, m$. We can apply quadrature methods like sparse grids or sampling techniques like (quasi) Monte-Carlo schemes, for example. Each method is defined by its nodes $\{\mathbf{p}_1, \ldots, \mathbf{p}_s\} \subset \Pi$ and weights $\{\gamma_1, \ldots, \gamma_s\} \subset \mathbb{R}$. We solve the linear dynamical systems

$$
\begin{aligned}
\mathbf{E}(\mathbf{p}_j)\dot{\mathbf{x}}(t, \mathbf{p}_j) &= \mathbf{A}(\mathbf{p}_j)\mathbf{x}(t, \mathbf{p}_j) + \mathbf{B}(\mathbf{p}_j)u(t) \\
y(t, \mathbf{p}_j) &= \mathbf{C}(\mathbf{p}_j)\mathbf{x}(t, \mathbf{p}_j)
\end{aligned}
$$
$$(19.11)$$

separately for $j = 1, \ldots, s$. The integrals in (19.4) change into the finite sums

$$
\hat{w}_i(t) := \sum_{j=1}^{s} \gamma_j \Phi_i(\mathbf{p}_j) y(t, \mathbf{p}_j) = \sum_{j=1}^{s} \gamma_j \Phi_i(\mathbf{p}_j) \mathbf{C}(\mathbf{p}_j)\mathbf{x}(t, \mathbf{p}_j)
$$
$$(19.12)$$

for $i = 1, \ldots, m$.

The systems (19.11) are merged into a single system as done in [20]. Let $\hat{\mathbf{x}}(t) := (\mathbf{x}(t, \mathbf{p}_1), \ldots, \mathbf{x}(t, \mathbf{p}_s)) \in \mathbb{R}^{ns}$ and $\hat{\mathbf{w}}(t) := (\hat{w}_1(t), \ldots, \hat{w}_m(t)) \in \mathbb{R}^m$. The systems (19.11) for $j = 1, \ldots, s$ together with the outputs (19.12) for $i = 1, \ldots, m$ yield the larger system

$$\hat{\mathbf{E}}\dot{\hat{\mathbf{x}}}(t) = \hat{\mathbf{A}}\hat{\mathbf{x}}(t) + \hat{\mathbf{B}}u(t)$$
$$\hat{\mathbf{w}}(t) = \hat{\mathbf{C}}\hat{\mathbf{x}}(t). \tag{19.13}$$

The system (19.13) consists of s separate subsystems (19.11), which are coupled just by the supply of the same input and the definition of the outputs (19.12). Thus the matrices $\hat{\mathbf{A}}, \hat{\mathbf{E}} \in \mathbb{R}^{ns \times ns}$ are block-diagonal. A column vector $\hat{\mathbf{B}} \in \mathbb{R}^{ns}$ is included. Obviously, the system (19.13) is asymptotically stable if the original systems (19.1) are asymptotically stable for all $\mathbf{p} \in \Pi$.

The approximations (19.12) yield the matrix $\hat{\mathbf{C}} \in \mathbb{R}^{m \times ns}$. We define the matrix

$$\mathbf{F} = (f_{ij}) \in \mathbb{R}^{m \times s}, \quad f_{ij} := \gamma_j \Phi_i(\mathbf{p}_j). \tag{19.14}$$

The row vector $\mathbf{C}$ in (19.1) is often independent of physical parameters. In this case, the larger matrix from (19.13) reads as

$$\hat{\mathbf{C}} = \mathbf{F}(\mathbf{I}_s \otimes \mathbf{C}) \tag{19.15}$$

with the identity matrix $\mathbf{I}_s \in \mathbb{R}^{s \times s}$ and the Kronecker product. The auxiliary system (19.13) is single-input-multiple-output (SIMO) now. The outputs $\hat{\mathbf{w}}$ yield the approximation (19.6) for the quantity of interest.

19.3.2 Symmetric Probability Distributions

An interesting property of sparse grid quadratures is that the matrix $\mathbf{F}$ from (19.14) exhibits many rows identical to zero in the case of polynomial bases and symmetric probability density functions (uniform distribution, Gaussian distribution, for example). The reason is that the multivariate polynomials are just the products of the univariate orthonormal polynomials. In the case of symmetric probability density functions, the univariate polynomials of odd degree are odd functions, see [9, p. 582]. Let $\overline{p}_\ell$ be the expected value of the ℓth random variable. It follows that

$$\phi_j^{(\ell)}(\overline{p}_\ell) = 0 \qquad \text{for odd } j.$$

A node of a sparse grid often features one or more components at a center point $\overline{p}_\ell$. Due to (19.7), the evaluation $\Phi_i(\mathbf{p}_j)$ becomes zero if at least one univariate polynomial results to zero. In the case of large numbers q of random parameters, the products consist of q terms and often some term is zero. This discussion also applies (with lower extend) to the case, where only a subset of the random variables have a symmetric probability distribution.

Consequently, the matrix $\hat{\mathbf{C}}$ in (19.15) has the same zero-rows as the matrix $\mathbf{F}$ from (19.14) both for constant vectors $\mathbf{C}$ and parameter-dependent vectors $\mathbf{C}(\mathbf{p})$ in (19.1). The approximation (19.6) becomes

$$\hat{\mathbf{y}}^{(\mathbb{I})}(t, \mathbf{p}) = \sum_{i \in \mathbb{I}} \hat{w}_i(t) \Phi_i(\mathbf{p}) = \sum_{i \in \mathbb{I} \setminus \mathbb{I}_0} \hat{w}_i(t) \Phi_i(\mathbf{p}),$$

where the subset $\mathbb{I}_0$ contains the indices of the zero-rows. Hence the quadrature method cancels out many output components in the sum (19.6) now, which already represents some sparsification. Since the sparse grid quadrature exhibits some accuracy, only sufficiently small outputs are approximated by zero and thus neglected. In the system (19.13), the number of outputs decreases from $m = |\mathbb{I}|$ to $|\mathbb{I} \setminus \mathbb{I}_0|$ by removing the zero-rows from the matrix $\hat{\mathbf{C}}$. This property represents an advantage for MOR, because many reduction methods suffer from a large number of outputs due to higher computational effort or convergence problems of iteration schemes.

Quadrature methods often exhibit a polynomial exactness, i.e., the formula yields the exact integrals for all multivariate polynomials up to a specific total degree d^*. The property discussed above does not contradict the polynomial exactness. Let d_i be the total degree of the ith basis polynomial. It holds that

$$\delta_{ik} = \ <\Phi_i, \Phi_k> \ = \sum_{j=1}^{s} \gamma_j \Phi_i(\mathbf{p}_j) \Phi_k(\mathbf{p}_j) \qquad \text{for} \quad d_i + d_k \leq d^*$$

with the Kronecker delta. If $d_i \leq \lfloor \frac{d^*}{2} \rfloor$, then the ith row of $\mathbf{F}$ cannot be identical to zero, because it follows that $<\Phi_i, \Phi_i> = 1$ and $d_i + d_i \leq d^*$. Consequently, zero-rows may appear only for basis polynomials with total degree $d_i > \lfloor \frac{d^*}{2} \rfloor$. For an index set (19.7) with fixed d, the phenomenon of zero rows disappears for sufficiently large d^*. However, such a high accuracy often cannot be afforded in the case of high-dimensional parameter sets Π.

19.4 Model Order Reduction

We use MOR for the construction of a low-dimensional approximation now.

19.4.1 Sparse Approximation by Model Order Reduction

The original system (19.1) may be relatively small. Nevertheless, the auxiliary system (19.13) becomes high-dimensional in the case of large numbers q of random parameters, because a quadrature method typically requires a high number s of nodes. Thus the linear dynamical system (19.13) allows for an MOR. A reduced order model (ROM) reads as

$$\overline{\mathbf{E}}\dot{\overline{\mathbf{x}}}(t) = \overline{\mathbf{A}}\overline{\mathbf{x}}(t) + \overline{\mathbf{B}}u(t)$$
$$\overline{\mathbf{w}}(t) = \overline{\mathbf{C}}\,\overline{\mathbf{x}}(t) \tag{19.16}$$

with dimensionality $r \ll ns$. Often projection-based MOR is used, where the matrices of the system (19.16) are defined by

$$\overline{\mathbf{A}} = \mathbf{T}_1^\top \hat{\mathbf{A}} \mathbf{T}_r, \quad \overline{\mathbf{B}} = \mathbf{T}_1^\top \hat{\mathbf{B}}, \quad \overline{\mathbf{C}} = \hat{\mathbf{C}} \mathbf{T}_r, \quad \overline{\mathbf{E}} = \mathbf{T}_1^\top \hat{\mathbf{E}} \mathbf{T}_r \tag{19.17}$$

with projection matrices $\mathbf{T}_1, \mathbf{T}_r \in \mathbb{R}^{ns \times r}$ of full rank. Established projection-based MOR methods for linear dynamical systems are moment matching techniques, balanced truncation and proper orthogonal decomposition, see [1, 11].

The output $\overline{\mathbf{w}}(t) := (\overline{w}_1(t), \ldots, \overline{w}_m(t))^\top \in \mathbb{R}^m$ of (19.16) yields an approximation $\hat{y}^{(\mathbb{I})}$ of the type (19.6) with $\hat{w}_i = \overline{w}_i$ for each i. We assume that the system (19.16) is asymptotically stable. In our MOR, a critical property of the system (19.13) is that the number of outputs is large for high-dimensional parameter sets Π.

As in [19], the crucial observation is

$$\overline{y}^{(r)}(t, \mathbf{p}) := \sum_{i=1}^m \overline{w}_i(t) \Phi_i(\mathbf{p}) = \sum_{i=1}^m \left[\sum_{j=1}^r \overline{c}_{ij} \overline{x}_j(t) \right] \Phi_i(\mathbf{p})$$
$$= \sum_{j=1}^r \overline{x}_j(t) \left[\sum_{i=1}^m \overline{c}_{ij} \Phi_i(\mathbf{p}) \right]. \tag{19.18}$$

If it holds that $r \ll m = |\mathbb{I}|$ with a sufficiently accurate ROM, then we achieve a sparse approximation of the type (19.9) with $\overline{v}_j = \overline{x}_j$ for each j. The formula (19.18) yields the alternative functions $\{\Psi_1, \ldots, \Psi_r\}$ given by

$$\Psi_j(\mathbf{p}) := \sum_{i=1}^m \overline{c}_{ij} \Phi_i(\mathbf{p}) \qquad \text{for } j = 1, \ldots, r. \tag{19.19}$$

In most of the cases, the system $\{\Psi_1, \ldots, \Psi_r\}$ is linearly independent. However, the basis functions (19.19) are not orthogonal. An orthonormal basis can be constructed by a singular value decomposition of the matrix $\overline{\mathbf{C}}$ from (19.16) as shown in [19, Sect. 4.2].

19.4.2   Numerical Rank Deficiency of Output Matrix

The output matrix may exhibit a (numerical) rank deficiency. The projection-based MOR (19.17) yields $\overline{\mathbf{C}} = \hat{\mathbf{C}} \mathbf{T}_r$. For a matrix-matrix-product, the rank fulfils the well-known inequalities

$$\text{rank}(\hat{\mathbf{C}}) + \text{rank}(\mathbf{T}_r) - ns \le \text{rank}(\overline{\mathbf{C}}) \le \min\{\text{rank}(\hat{\mathbf{C}}), \text{rank}(\mathbf{T}_r)\}. \qquad (19.20)$$

The projection matrix $\mathbf{T}_r$ has full rank r. We assume that the output matrix $\hat{\mathbf{C}}$ owns full rank m, which is often satisfied. In view of $m \ll ns$ as well as $r \ll ns$, the lower bound in (19.20) becomes trivial. The upper bound of (19.20) implies $\text{rank}(\overline{\mathbf{C}}) \le r$ due to $r < m$. Since the maximum rank of the output matrix would be r, a rank deficiency is not excluded.

Numerical simulations show a numerical rank deficiency of the matrix, which sounds surprising. The matrix $\hat{\mathbf{C}} \in \mathbb{R}^{m \times ns}$ shrinks to $\overline{\mathbf{C}} \in \mathbb{R}^{m \times r}$ with $r \ll ns$. Thus we expect that all columns of $\overline{\mathbf{C}}$ include important information. Yet not all MOR methods are optimal in this sense, i.e., also unessential parts may appear.

If a rank deficiency or numerical rank deficiency of the output matrix occurs, then it can be removed using the singular value decomposition of $\overline{\mathbf{C}}$ again as discussed in [19, Sect. 4.2]. Keeping all singular values above the machine precision avoids a (numerical) rank deficiency. A restriction to the dominant singular values allows for a further reduction of the dimensionality in the sparse approximation (19.18) from r to some r'. However, still the ROM (19.16) of dimension r has to be solved to obtain an approximation of dimension $r' < r$.

19.4.3   Error Estimates

The input-output behaviour of a linear time-invariant dynamical system is described by a transfer function in the frequency domain, see [1, 11]. The coupled system (19.13) owns a transfer function $\hat{H} : \hat{S} \to \mathbb{C}^m$ with $\hat{S} \subset \mathbb{C}$. Likewise, the reduced system (19.16) has a transfer function $\overline{H} : \overline{S} \to \mathbb{C}^m$ with $\overline{S} \subset \mathbb{C}$. It holds that $\hat{S}$ and $\overline{S}$ include the imaginary axis provided that the two systems are asymptotically stable. For a transfer function G of an SISO system, the Hardy norms $\mathscr{H}_\infty$ and $\mathscr{H}_2$ are defined by

$$\|G\|_{\mathscr{H}_\infty} = \sup_{\omega \in \mathbb{R}} |G(\mathrm{i}\,\omega)| \qquad \text{and} \qquad \|G\|_{\mathscr{H}_2} = \sqrt{\frac{1}{2\pi} \int_{-\infty}^{+\infty} |G(\mathrm{i}\,\omega)|^2 \, \mathrm{d}\omega}$$

$$(19.21)$$

with $\mathrm{i} := \sqrt{-1}$. We employ these Hardy norms component-wise to $\hat{H}$ and $\overline{H}$.

Let the initial values of the systems (19.1) be identical to zero. The difference between the quantity of interest from the coupled system (19.13) and the quantity of interest from the reduced system (19.16) satisfies the estimates

$$\sup_{t \geq 0} \left\| \hat{y}^{(\mathbb{I})}(t, \cdot) - \overline{y}^{(r)}(t, \cdot) \right\|_{\mathscr{L}^2(\Pi, \rho)} \leq \sqrt{ \sum_{i=1}^{m} \left\| \hat{H}_i - \overline{H}_i \right\|_{\mathscr{H}_2}^2 } \; \|u\|_{\mathscr{L}^2[0, \infty)} \quad (19.22)$$

$$\left\| \hat{y}^{(\mathbb{I})} - \overline{y}^{(r)} \right\|_{\mathscr{L}^2(\Pi, \rho) \times \mathscr{L}^2[0, \infty)} \leq \sqrt{ \sum_{i=1}^{m} \left\| \hat{H}_i - \overline{H}_i \right\|_{\mathscr{H}_\infty}^2 } \; \|u\|_{\mathscr{L}^2[0, \infty)} \quad (19.23)$$

provided that the norms are finite. Therein, the $\mathscr{L}^2$-norm of the time domain appears for the input. The proof can be obtained as in [19, Thm. 1]. The error bounds (19.22) and (19.23) are not sharp and thus the true approximation error may be much smaller. The sparsification error in (19.10) coincides with the error of this MOR.

19.5   Illustrative Example

We apply the approach from the previous sections now. The following computations were done using the software package MATLAB [16].

In [20], an ODE model of an L-C-Π band pass filter from [15] was investigated including $q = 11$ physical parameters. Now we extend this band pass filter to a sequence of components shown in Fig. 19.1. An input voltage is added, whereas the output voltage drops at a load resistance. We choose six components resulting in $q = 41$ physical parameters: 13 capacitances, 13 inductances and 15 resistances. We derive a linear system (19.1) of DAEs with $n = 41$ equations for 27 node voltages and 14 branch currents by modified nodal analysis, see [13]. This DAE system owns the nilpotency index one for all physically relevant parameters. Moreover, the system is asymptotically stable as well as strictly proper. Figure 19.2 depicts the Bode diagram of the linear dynamical system for a constant choice of the parameters: $C = 10^{-6}$ for all capacitances, $L = 10^{-6}$ for all inductances and $R = 0.1$ for all resistances except for $R_{\text{in}} = 1$ at the input and $R_{\text{out}} = 10^3$ at the output. The magnitude of the transfer function shows that frequencies outside a relatively small interval are damped strongly.

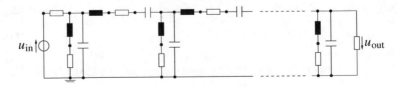

Fig. 19.1 Electric circuit of an L-C-Π band pass filter

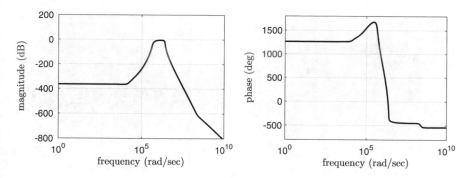

Fig. 19.2 Bode plot for transfer function of band pass filter in the case of deterministic parameters

Concerning the stochastic modelling, we introduce independent uniform distributions for each parameter, which describe a perturbation of up to 20% around the constant parameter choice from above. In the truncated expansion (19.5), we include all multivariate polynomials up to total degree $d = 3$. The number of basis polynomials becomes $m = 13,244$ due to (19.8).

We use two methods for the approximation of the coefficient functions (19.12):

1. a sparse grid by Smolyak construction of level two based on the one-dimensional Clenshaw-Curtis rule from [7]. The number of nodes is $s = 3445$. Therein, negative weights appear. The degree of polynomial exactness is five.
2. a quasi Monte-Carlo method using a Sobol sequence from the built-in routine sobolset of MATLAB [16]. We apply $s = 5000$ samples. The weights are all positive and identical. The scheme does not exhibit a polynomial exactness.

We arrange the coupled system (19.13) for each method. The dimensions result in $ns = 141,245$ and $ns = 205,000$, respectively. The number of outputs is identical to m in both systems.

An approximation of the Hardy norms (19.21) is computed using a logarithmically spaced grid in the frequency interval $\omega \in [1, 10^{10}]$ on the imaginary axis. The $\mathcal{H}_\infty$-norms of the systems (19.13) are depicted in Fig. 19.3. Both norms $\mathcal{H}_\infty$ and $\mathcal{H}_2$ are shown by Fig. 19.4 in descending order. We recognise a faster decay of the norms for the sparse grid quadrature. Observing the sparse grid technique, about 80.5% of the components exhibit norms identical to zero, because the associated rows of the matrix $\hat{C}$ in (19.13) are zero due to the phenomenon described in Sect. 19.3.2. In contrast, the norms are all non-zero for the quasi Monte-Carlo scheme. The faster decay indicates a higher accuracy of the quadrature method, since the exact norms often decrease rapidly.

In the MOR, we use the Arnoldi procedure, which represents a simple moment matching technique using Krylov subspaces, see [11]. The single expansion point $\mu = 10^7$ is applied for the moment matching. Thus the matrices in the ROM (19.16),(19.17) become real-valued. An arbitrary dimension $r < ns$ of

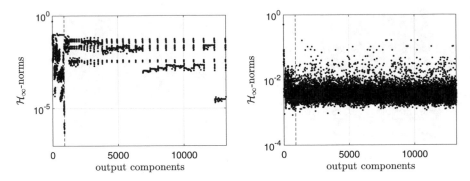

Fig. 19.3 $\mathcal{H}_\infty$-norms of the system (19.13) for the sparse grid quadrature (left) and the quasi Monte-Carlo method (right). The dashed line separates components for polynomials of degree lower than three and degree three (semi-logarithmic scale. Values identical to zero are ignored)

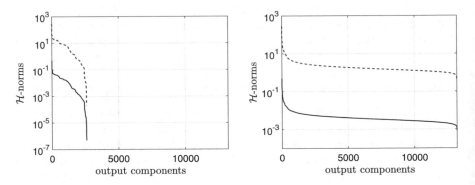

Fig. 19.4 Hardy norms of the system (19.13) for the sparse grid quadrature (left) and the quasi Monte-Carlo method (right) in descending order: $\mathcal{H}_\infty$-norms (solid lines) and $\mathcal{H}_2$-norms (dashed lines) (semi-logarithmic scale. Values identical to zero are ignored)

the ROM can be chosen. In our context, a crucial advantage of the Arnoldi procedure is that the computational effort is nearly independent of the number of outputs. However, a disadvantage is that the reduced systems (19.16) are not necessarily stable.

We perform the MOR for dimensions $r = 10, 20, 30, \ldots, 400$. Figure 19.5 illustrates the error bounds (19.22) and (19.23) given a unit norm of the input with the selected dimensions of the reduced system (19.16). We observe that the error estimates decrease significantly for larger dimensions. Both the magnitude and the speed of decay are similar for both methods. From the 40 reduced systems, just 12 and 25 systems are stable for the sparse grid quadrature and the quasi Monte-Carlo method, respectively. The instabilities occur for both smaller and larger reduced dimensions. Furthermore, Table 19.1 shows the minimum dimensions r to obtain an ROM, where the error bound is below a threshold $\delta > 0$ and the reduced system is stable. The sparsity for $r = 400$ becomes $\sigma = 0.03$, for example, assuming that polynomials of total degree $d = 3$ are required for a sufficiently small truncation error.

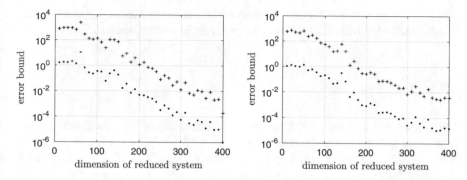

Fig. 19.5 Error bounds (19.22) (+) and (19.23) (·) for sparse grid quadrature (left) and quasi Monte-Carlo method (right) depending on the dimension of the reduced system (semi-logarithmic scale)

Table 19.1 Minimum dimension $r \in \{10, 20, 30, \ldots, 400\}$ required for an error bound below a threshold δ together with a stable reduced system

Threshold δ		10^{-1}	10^{-2}	10^{-3}	10^{-4}	10^{-5}
Sparse grid	Error bound (19.22)	250	320	400	–	–
	Error bound (19.23)	120	180	250	300	400
Quasi M.-C.	Error bound (19.22)	230	300	–	–	–
	Error bound (19.23)	160	160	190	280	–

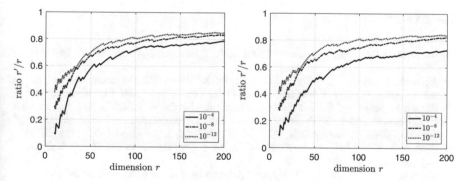

Fig. 19.6 Ratios r'/r between reduced dimension r and additional reduction to r' by neglecting singular values of the output matrix below the thresholds $10^{-4}, 10^{-8}, 10^{-12}$ for sparse grid quadrature (left) and quasi Monte-Carlo method (right)

Finally, we investigate the singular value decomposition of the output matrix for the ROM (19.16) as discussed in Sect. 19.4.2. The dimension r of the sparse approximation (19.18) can be reduced further to some r' by omitting all singular values below a threshold, which has to be larger or equal than the machine precision. The resulting ratios are depicted for different reduced dimensions and miscellaneous thresholds in Fig. 19.6. We recognise that the behaviour is similar in both stochastic collocation methods. For higher dimensions r, the additional reduction can decrease the dimension by about 20% without a significant loss of accuracy.

19.6 Conclusions

We constructed a sparse approximation for a quantity of interest given a linear dynamical system including random variables. The approximation requires just a few basis functions depending on the random variables. The construction is feasible using an arbitrary quadrature method or sampling technique and an arbitrary MOR method for large linear dynamical systems. Both the general analysis and the numerical simulation of a test example demonstrate a promising potential for the efficient determination of a low-dimensional approximation. Furthermore, the performance of a sparse grid quadrature was better than a quasi Monte-Carlo method in the test example. In the sparse grid variant, the Hardy norms of the outputs decrease faster, which reproduces the behaviour of the exact outputs. The efficiency can be improved by the application of MOR methods, which are stability-preserving, see [6], as well as powerful in the case of large numbers of outputs, see [3].

References

1. Antoulas, A.C.: Approximation of Large-Scale Dynamical Systems. SIAM, Philadelphia (2005)
2. Augustin, F., Gilg, A., Paffrath, M., Rentrop, P., Wever, U.: Polynomial chaos for the approximation of uncertainties: chances and limits. Eur. J. Appl. Math. **19**, 149–190 (2008)
3. Benner, P., Schneider, A.: Balanced truncation model order reduction for LTI systems with many inputs or outputs. In: Proceedings of the 19th International Symposium on Mathematical Theory of Networks and Systems, MTNS 2010, Budapest, July 5–9, pp. 1971–1974 (2010)
4. Benner, P., Hinze, M., ter Maten, E.J.W. (eds.): Model Reduction for Circuit Simulation. Springer, Dordrecht (2011)
5. Blatman, G., Sudret, B.: Adaptive sparse polynomial chaos expansion based on least angle regression. J. Comput. Phys. **230**, 2345–2367 (2011)
6. Bond, B.N.: Stability-preserving model reduction for linear and nonlinear systems arising in analog circuit applications. Ph.D. thesis, Massachusetts Institute of Technology (2010)
7. Burkardt, J.: Sparse grids based on the Clenshaw Curtis rule. Online document (2009). http://people.sc.fsu.edu/~jburkardt/m_scr/sparse_grid_cc/sparse_grid_cc.html [cited 14 Sept 2016]
8. Conrad, P.R., Marzouk, Y.M.: Adaptive Smolyak pseudospectral approximations. SIAM J. Sci. Comput. **35**, A2643–A2670 (2013)
9. Dahlquist, G., Björck, A.: Numerical Methods in Scientific Computing, vol. I. SIAM, Philadelphia (2008)
10. Doostan, A., Owhadi, H.: A non-adapted sparse approximation of PDEs with stochastic inputs. J. Comput. Phys. **230**, 3015–3034 (2011)
11. Freund, R.: Model reduction methods based on Krylov subspaces. Acta Numer. **12**, 267–319 (2003)

12. Gerstner, T., Griebel, M.: Numerical integration using sparse grids. Numer. Algorithms **18**, 209–232 (1998)
13. Ho, C.W., Ruehli, A., Brennan, P.: The modified nodal approach to network analysis. IEEE Trans. Circuits Syst. **22**, 504–509 (1975)
14. Jakeman, J.D., Eldred, M.S., Sargsyan, K.: Enhancing ℓ_1-minimization estimates of polynomial chaos expansions using basis selection. J. Comput. Phys. **289**, 18–34 (2015)
15. Kessler, R.: Aufstellen und numerisches Lösen von Differential-Gleichungen zur Berechnung des Zeitverhaltens elektrischer Schaltungen bei beliebigen Eingangs-Signalen. Online document (2007). http://www.home.hs-karlsruhe.de/~kero0001/aufst6/AufstDGL6hs.html [cited 22 Aug 2016]
16. MATLAB, version 8.6.0 (R2015b). The Mathworks Inc., Natick (2015)
17. Niederreiter, H.: Random Number Generation and Quasi-Monte Carlo Methods. SIAM, Philadelphia (1992)
18. Pulch, R.: Stochastic collocation and stochastic Galerkin methods for linear differential algebraic equations. J. Comput. Appl. Math. **262**, 281–291 (2014)
19. Pulch, R.: Model order reduction and low-dimensional representations for random linear dynamical systems. Math. Comput. Simulat. **144**, 1–20 (2018)
20. Pulch, R.: Model order reduction for stochastic expansions of electric circuits. In: Bartel, A., Clemens, M., Günther, M., ter Maten, E.J.W. (eds.) Scientific Computing in Electrical Engineering SCEE 2014. Mathematics in Industry, vol. 23, pp. 223–231. Springer, Berlin (2016)
21. Pulch, R., ter Maten, E.J.W.: Stochastic Galerkin methods and model order reduction for linear dynamical systems. Int. J. Uncertain. Quantif. **5**, 255–273 (2015)
22. Pulch, R., ter Maten, E.J.W., Augustin, F.: Sensitivity analysis and model order reduction for random linear dynamical systems. Math. Comput. Simul. **111**, 80–95 (2015)
23. Stroud, A.: Approximate Calculation of Multiple Integrals. Prentice Hall, Upper Saddle River (1971)
24. Xiu, D.: Numerical Methods for Stochastic Computations: A Spectral Method Approach. Princeton University Press, Princeton (2010)

Chapter 20
POD-Based Reduced-Order Model of an Eddy-Current Levitation Problem

Md. Rokibul Hasan, Laurent Montier, Thomas Henneron, and Ruth V. Sabariego

Abstract The accurate and efficient treatment of eddy-current problems with movement is still a challenge. Very few works applying reduced-order models are available in the literature. In this paper, we propose a proper-orthogonal-decomposition reduced-order model to handle these kind of motional problems. A classical magnetodynamic finite element formulation based on the magnetic vector potential is used as reference and to build up the reduced models. Two approaches are proposed. The TEAM workshop problem 28 is chosen as a test case for validation. Results are compared in terms of accuracy and computational cost.

20.1 Introduction

The finite element (FE) method is widely used and versatile for accurately modelling electromagnetic devices accounting for eddy current effects, non-linearities, movement,... However, the FE discretization may result in a large number of unknowns, which maybe extremely expensive in terms of computational time and memory. Furthermore, the modelling of a movement requires either remeshing or ad-hoc techniques. Without being exhaustive, it is worth mentioning: the hybrid finite-element boundary-element (FE-BE) approaches [1], the sliding mesh techniques (rotating machines) [2] or the mortar FE approaches [3].

Md. R. Hasan (✉) · R. V. Sabariego
KU Leuven, Dept. Electrical Engineering (ESAT), Leuven, Belgium

EnergyVille, Genk, Belgium
e-mail: rokib.hasan@esat.kuleuven.be; ruth.sabariego@esat.kuleuven.be

L. Montier · T. Henneron
Laboratoire d'Electrotechnique et d'Electronique de Puissance, Arts et Metiers ParisTech,
Lille, France
e-mail: laurent.montier@ensam.eu; thomas.henneron@univ-lille1.fr

© Springer International Publishing AG, part of Springer Nature 2018
U. Langer et al. (eds.), *Scientific Computing in Electrical Engineering*,
Mathematics in Industry 28, https://doi.org/10.1007/978-3-319-75538-0_20

Physically-based reduced models are the most popular approaches for efficiently handling these issues. They extract physical parameters (inductances, flux linkages, . . .) either from simulations or measurements and construct look-up tables covering the operating range of the device at hand [4, 5]. Future simulations are performed by simple interpolation, drastically reducing thus the computational cost. However, these methods depend highly on the expert's knowledge to choose and extract the most suitable parameters.

Mathematically-based reduced-order (RO) techniques are a feasible alternative, which are gaining interest in electromagnetism [6]. RO modelling of static coupled system has already been implemented in [7, 8]. Few RO works have addressed problems with movement (actuators, electrical machines, etc.) [9–11].

In [9], authors consider a POD-based FE-BE model electromagnetic device comprising nonlinear materials and movement. Meshing issues are avoided but the system matrix is not sparse any more, increasing considerably the cost of generating the RO model. In [10], a magnetostatic POD-RO model of a permanent magnet synchronous machine is studied. A locked step approach is used, so the mesh and associated number of unknowns remains constant. A POD-based block-RO model is proposed in [11, 12], where the domain is split in linear and nonlinear regions and the ROM is applied only to the linear part.

In this paper, we consider a POD-based FE model of a levitation problem, namely the Team Workshop problem 28 (TWP28) [5, 13] (a conducting plate above two concentric coils, see Fig. 20.1). The movement is modelled with two RO models based on: (1) FE with automatic remeshing of the complete domain; (2) FE with constraint remeshing, i.e., localized deformation of the mesh around the moving plate, hereafter referred to as mesh deformation. Both models are validated in the time domain and compared in terms of computational efficiency.

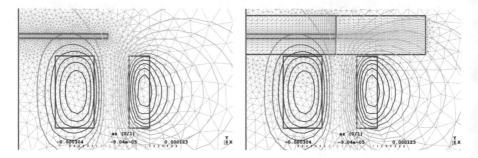

Fig. 20.1 2D axisymmetric mesh of TWP28: aluminium plate above two concentric coils (12.8 mm clearance). Real part of the magnetic flux density. Left: automatic remeshing of the full domain, Right: mesh deformation of sub-domain around plate with nodes fixed at it's boundaries (except axes)

20.2   Magnetodynamic Levitation Model

Let us consider a bounded domain $\Omega = \Omega_c \cup \Omega_c^C \in \mathbb{R}^3$ with boundary Γ. The conducting and non-conducting parts of Ω are denoted by Ω_c and Ω_c^C, respectively. The (modified) magnetic-vector-potential $(a$-$)$ magnetodynamic formulation (weak form of Ampère's law) reads: find a, such that

$$(\nu \operatorname{curl} a, \operatorname{curl} a')_\Omega + (\sigma \partial_t a, a')_{\Omega_c} + \langle \hat{n} \times h, a'\rangle_\Gamma = (j_s, a')_{\Omega_s}, \quad \forall a' \qquad (20.1)$$

with a' test functions in a suitable function space; $b(t) = \operatorname{curl} a(t)$, the magnetic flux density; $j_s(t)$ a prescribed current density and $\hat{n}$ the outward unit normal vector on Γ. Volume integrals in Ω and surface integrals on Γ of the scalar product of their arguments are denoted by $(\cdot, \cdot)_\Omega$ and $\langle \cdot, \cdot \rangle_\Omega$. The derivative with respect to time is denoted by ∂_t. We further assume linear isotropic and time independent materials with magnetic constitutive law, so that the magnetic field is $h(t) = \nu b(t)$ (reluctivity ν) and electric constitutive law, given by induced eddy current density $j(t) = \sigma e(t)$, (conductivity σ) where, electric field $e(t) = -\partial_t a(t)$. Assuming a rigid Ω_c (no deformation) and a purely translational movement (no rotation, no tilting), the electromagnetic force appearing due to the eddy currents in Ω_c can be modelled as a global quantity with only one component (vertical to the plate). If Ω_c is non-magnetic, Lorentz force can be used:

$$F_{em}(t) = \int_{\Omega_c} j(t) \times b(t) \, d\Omega_c = \int_{\Omega_c} -\sigma \partial_t a(t) \times \operatorname{curl} a(t) \, d\Omega_c . \qquad (20.2)$$

The 1D mechanical equation governing the above described levitation problem reads:

$$m \, \partial_t v(t) + \xi v(t) + k y(t) + mg = F_{em}(t) \qquad (20.3)$$

where unknown $y(t)$ is the center position of the moving body in the vertical direction, $v(t) = \partial_t y(t)$ is the velocity of the moving body, m is the mass of the moving body, g is the acceleration of gravity, ξ is the scalar viscous friction coefficient, k is the elastic constant. We apply the backward Euler method to solve (20.3). The moving body displacement of system (20.3) results from the ensuing electromagnetic force generated by system (20.1) and thus affects the geometry. Given that, the dynamics of the mechanical equation is much slower than the electromagnetic equation, if the time-step is taken sufficiently small, one can decouple the equations. Under this condition, the electromagnetic and mechanical equations can be solved alternatively rather than simultaneously by the weak electromechanical coupling algorithm of [14]. We adopt this approach.

20.3 POD-Based Model Order Reduction

The proper orthogonal decomposition (POD) is applied to reduce the matrix system resulting from the FE discretisation of (20.1):

$$A \partial_t x(t) + B x(t) = C(t) . \tag{20.4}$$

where $x(t) \in \mathbb{R}^{N \times 1}$ is the time-dependent column vector of N unknowns, $A, B \in \mathbb{R}^{N \times N}$ are the matrices of coefficients and $C(t) \in \mathbb{R}^{N \times 1}$ is the source column vector. Furthermore, the system (20.4) is discretized in time by means of the backward Euler scheme. A system of algebraic equation is obtained for each time step from t_{k-1} to $t_k = t_{k-1} + \Delta t$, Δt the step size. The discretized system reads:

$$[A_{\Delta t} + B] x_k = A_{\Delta t} x_{k-1} + C_k \tag{20.5}$$

with $A_{\Delta t} = \frac{A}{\Delta t}$, $x_k = x(t_k)$ the solution at instant t_k, $x_{k-1} = x(t_{k-1})$ the solution at instant t_{k-1}, C_k the right-hand side at instant t_k.

In RO techniques, the solution vector $x(t)$ is approximated by a vector $x^r(t) \in \mathbb{R}^{M \times 1}$ within a reduced subspace spanned by $\Psi \in \mathbb{R}^{N \times M}$, $M \ll N$,

$$x(t) \approx \Psi x^r(t) , \tag{20.6}$$

with Ψ an orthonormal projection operator generated from the time-domain full solution $x(t)$ via snapshot techniques [15].

Let us consider the snapshot matrix, $S = [x_1, x_2, \ldots, x_M] \in \mathbb{R}^{N \times M}$ from the set of solution x_k for the selected number of time steps. Applying the singular value decomposition (SVD) to S as,

$$S = \mathcal{U} \Sigma \mathcal{V}^T . \tag{20.7}$$

where Σ contains the singular values, ordered as $\sigma_1 > \sigma_2 > \ldots > 0$. We consider $\Psi = \mathcal{U}^r \in \mathbb{R}^{N \times r}$, that corresponds to the truncation (r first columns, which has larger singular values than a pre-defined error tolerance ε) with orthogonal matrices $\mathcal{U} \in \mathbb{R}^{N \times r}$ and $\mathcal{V} \in \mathbb{R}^{M \times r}$. Therefore, the RO system of (20.5) reads

$$\left[A^r_{\Delta t} + B^r \right] x^r_k = A^r_{\Delta t} x^r_{k-1} + C^r_k , \tag{20.8}$$

with $A^r_{\Delta t} = \Psi^T A_{\Delta t} \Psi$, $B^r = \Psi^T B \Psi$ and $C^r = \Psi^T C$ [16].

20.3.1   Application to an Electro-Mechanical Problem with Movement

20.3.1.1   RO Modelling with Automatic Remeshing Technique

In case of automatic remeshing, we transfer results from the source $mesh_{k-1}$ to the new target $mesh_k$ by means of a Galerkin projection, which is optimal in the L_2-norm sense [17]. Note that, this projection is limited to the conducting domain, i.e. the plate, as it is only there that we need to compute the time derivative. The number of unknowns per time step t_k varies and the construction of the snapshot matrix S is not straightforward. As the solution at t_k is supported on its own mesh, the snapshot vectors x_k have a different size. They have to be projected to a common basis using a simple linear interpolation technique before being assembled in S and getting the projection operator Ψ. The procedure becomes thus extremely inefficient.

20.3.1.2   RO Modelling with Mesh Deformation Technique

The automatic remeshing task is replaced by a mesh deformation technique, limited to a region around the moving body (see, e.g., the box in Fig. 20.2). Therefore, in this case, the remeshing is done by deforming the initial mesh, which is generated with the conducting plate placed at, e.g., y_0 (avoiding bad quality elements), see Fig. 20.2. The mesh elements only inside the sub-domain can be deformed (shrink/expand), see Fig. 20.3 and the nodes at the boundary of the sub-domain are fixed. The

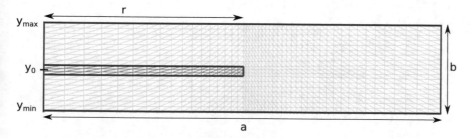

Fig. 20.2 Sub-domain for deformation: plate position at $y_0 = 12.8$ mm (initial mesh)

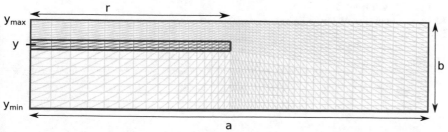

Fig. 20.3 Sub-domain for deformation: plate position at $y = 20$ mm. Mesh elements under the plate are expanded and above the plate are shrinked

surrounding mesh does not vary. In our test case, we assume a vertical force (neglect the other two components) in (20.2), therefore, the mesh elements only deform in the vertical direction and the nodes are fixed at the boundary of the sub-domain (not at the axes due to the axisymmetry). The size of the sub-domain (a×b) is determined by the extreme positions of the moving body. In our validation example, the minimum position (3.8 mm) is given by the upper borders of the coils and the maximum position (22.3 mm) could be estimated by means of a circuital model, e.g. [5]. The number of unknowns per time step remains now constant so the construction of matrix S is direct.

Algorithm 1: Automatic remeshing

Input : snapshot vectors $\{S_c\} \leftarrow \{x_k\} \in \mathbb{R}^{n(k) \times 1}$
time steps
$\{t_k\}, k \in [1, \ldots, K]$
$A_{\Delta t}$, B, C, tolerance ε
$m \leq n(k)$ snapshot vectors
Output: displacement y_k

1 $y_0 = $ initial position, $\Delta y_0 = 0$

   //Time resolution
2 **for** $k \leftarrow 1$ **to** K **do**
      //Magnetics
3   generate matrices $A_{\Delta t_k}$, B_k, C_k
4   find length of $C_k \in \mathbb{R}^{n(k) \times 1}$
5       $S_p = \mathbf{0} \in \mathbb{R}^{n(k) \times m}$
6       $S_p \leftarrow$ projection of $\{S_c\}$ to $n(k)$ rank subspace
7   SVD of $S_p = \mathscr{U} \Sigma \mathscr{V}^T$
8       $\Psi_k = \mathscr{U}(:, 1 \ldots r)$ with r such that $\sigma(i)/\sigma(1) > \varepsilon, \forall i \in [1 \ldots r]$
      $A^r_{\Delta t_k} = \Psi_k^T A_{\Delta t_k} \Psi_k,$
9       $B^r_k = \Psi_k^T B_k \Psi_k,$
      $C^r_k = \Psi_k^T C_k$
10   solve $\left(A^r_{\Delta t_k} + B^r_k \right) x^r_k = C^r_k + A^r_{\Delta t_k} x^r_{k-1}$
11     $x_k \approx \Psi_k x^r_k$
12   compute force F_k
      //Mechanics
13   compute displacement y_k
14   update $\Delta y_k = y_k - y_{k-1}$
15   remesh with y_k
16 **end**

Algorithm 2: Mesh deformation

Input : snapshot matrix $S = [x_1, \ldots, x_m] \in \mathbb{R}^{n \times m}$, $x_k \in \mathbb{R}^{n \times 1}$
         time steps
         $\{t_k\}, k \in [1, \ldots, K]$
         $A_{\Delta t}$, B, C, tolerance ε
         $m \leq n$ snapshot vectors
Output: displacement y_k

1 $y_0 =$ initial position, $\Delta y_0 = 0$
2 get initial mesh
3 SVD of $S = \mathcal{U} \Sigma \mathcal{V}^T$
4 $\Psi_k = \mathcal{U}(:, 1 \ldots r)$ with r such that $\sigma(i)/\sigma(1) > \varepsilon, \forall i \in [1 \ldots r]$

   //Time resolution
5 **for** $k \leftarrow 1$ **to** K **do**
       //Magnetics
6   generate matrices $A_{\Delta t_k}$, B_k, C_k
       $A^r_{\Delta t_k} = \Psi^T A_{\Delta t_k} \Psi,$

7      $B^r_k = \Psi^T B_k \Psi,$

       $C^r_k = \Psi^T C_k$
8   solve $\left(A^r_{\Delta t_k} + B^r_k \right) x^r_k = C^r_k + A^r_{\Delta t_k} x^r_{k-1}$
9      $x_k \approx \Psi x^r_k$
10   compute force F_k

       //Mechanics
11   compute displacement y_k
12   update $\Delta y_k = y_k - y_{k-1}$
13   deform mesh with y_k
14 **end**

20.4 Application Example

We consider TWP28: an electrodynamic levitation device consisting of a conducting cylindrical aluminium plate ($\sigma = 3.47 \times 10^7$ S/m, $m = 0.107$ kg, $\xi = 1$) above two coaxial exciting coils. The inner and outer coils have 960 and 576 turns respectively. Note that, if we neglect the elastic force, the equilibrium is reached when the F_{em} is 1N. At $t = 0$, the plate rests above the coils at a distance of 3.8 mm. For $t \geq 0$, a time-varying sinusoidal current (20 A, $f = 50$ Hz) is imposed, same amplitude, opposite directions [13]. Assuming a translational movement (no rotation and tilting) we can use an axisymmetric model. A FE model is generated as reference and origin of the RO models. We have time-stepped 50 periods

(100 time steps per period and step size 0.2 ms), discretization that ensures accuracy and avoids degenerated mesh elements during deformation.

20.4.1 RO Modelling with Automatic Remeshing Full Domain

In case of full domain remeshing, the first 1500 time steps (300 ms) of the simulation, that correspond to the first two peaks (2P) in Fig. 20.4, are included in the snapshot matrix.

Three POD-based RO models are constructed based on the r number of first singular value modes greater than a prescribed error tolerance ε, that is set manually observing the singular values decay curve of the snapshot matrix (see in Table 20.1). The smaller the prescribed ε, the bigger the size of the RO model will be (size of RO3 > RO2 > RO1).

The displacement and relative error of the full and RO models are shown in Fig. 20.4. Accurate results have been achieved with the truncated basis models: RO2 and RO3, with fix size per time step $M = 1403$ and 1411. This approach is completely inefficient, as the maximum number of unknowns we have in the full model is 1552.

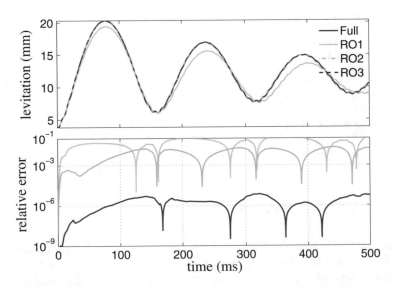

Fig. 20.4 Displacement (up) and relative error (down) between full and RO models

Table 20.1 L_2-relative errors of RO models on levitation height for 2P (automatic remeshing)

RO models	M	ε	Rel. error
RO1	1085	10^{-6}	1.25×10^{-1}
RO2	1403	10^{-11}	1.03×10^{-2}
RO3	1411	10^{-15}	2.45×10^{-6}

20.4.2   RO Modelling with Mesh Deformation of a Sub-domain

The choice of the sub-domain to deform the mesh is a non-trivial task: it should be as small as possible while ensuring a high accuracy. From our reference FE solution [13], by observing the minimum and maximum levitation height of the plate, we fixed the sub-domain size along the y-axis between $y_{min} = 1.3$ mm and $y_{max} = 29.3$ mm, distances measured from the upper border of the coils. The size along the x-axis has a minimum equal to the radius of the plate, i.e. $r = 65$ mm. This value is however not enough due to fringing effects. We have taken different size along the x-axis: $1.5r, 2r, 3r$ (97.5, 130 and 195 mm), measured from the axis (Fig. 20.2). The meshed boxes yield 1921, 1836 and 1780 number of unknowns.

The relative errors in time shown in Table 20.2 decrease with the increasing sub-domain lengths/box sizes considered. We have therefore chosen to further analyse the RO results obtained with a box length along x of 195 mm ($3r$). The discretization is kept constant for all RO models computation.

The first 800 time steps (160 ms) of the simulation, that correspond to the first peak (1P), are taken in the snapshot matrix in order to generate the projection basis Ψ. In the snapshot matrix, the most important time step solutions are included, which found as optimum selection for approximating the full solution. Then the basis is truncated as $\Psi = \mathscr{U}^r$ (r first columns) by means of prescribed error tolerance ($\varepsilon = 10^{-5}, 10^{-8}$). The basis are truncated for 1P to get RO models of size $M = 7$ and 35.

From Fig. 20.5, it can be observed that, RO model already shows very good argument with only $M = 7$ truncated basis, which is generated from the snapshot matrix that incorporates first peak (1P). The accuracy of RO models does not improve significantly with the addition of following transient peaks (2P) into the snapshot matrix, but the accuracy certainly improves with M. Hence, with $M = 35$ the full and RO curves are indistinguishable. The accuracy of the RO models can also be observed from the L_2-relative errors figure.

With regard to the computation time (5000 time steps), the RO with $M = 7$, can be solved less than an hour, which is 3.5 times faster than the full-domain automatic remeshing approach, where the major time consuming part is to project the Ψ on a same dimensional basis as the system coefficient matrices, to reduce the system in each time step. Be aware that the computation is not optimized, performed on a laptop, (Intel Core i7-4600U CPU at 2.10 GHz) without any parallelization.

Table 20.2 L_2-relative errors of RO models on levitation height for 1P (mesh deform)

Sub-domain lengths (mm)	$M = 7$	$M = 35$
97.5	8.24×10^{-2}	6.14×10^{-4}
130	5.71×10^{-2}	1.90×10^{-4}
195	4.53×10^{-3}	3.73×10^{-5}

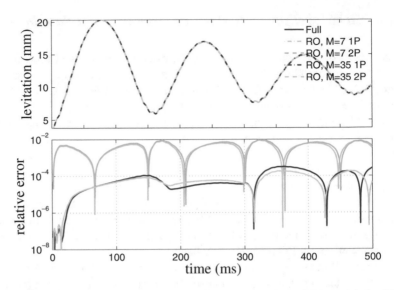

Fig. 20.5 Displacement (up) and relative error (down) between full and RO models for 195 mm sub-domain length

20.5 Conclusion

In this paper, we have proposed two approaches for POD-based RO models to treat a magnetodynamic levitation problem: automatic remeshing and mesh deformation of a sub-domain around a moving body. The RO model is completely inefficient with automatic remeshing technique, as the computational cost is nearly expensive as the classical approach. The approach with sub-domain deformation to limit the influence of the movement on the RO model construction has proved accurate and efficient (low computational cost). We have shown results for three different sub-domain sizes, the bigger the sub-domain the higher the accuracy. Further, computationally efficient RO modelling of such parametric model is ongoing research.

References

1. Sabariego, R., Gyselinck, J., Dular, P., Geuzaine, C., Legros, W.: Fast multipole acceleration of the hybrid finite-element/boundary-element analysis of 3-D eddy-current problems. IEEE Trans. Magn. **40**(2), 1278–1281 (2004)
2. Boualem, B., Piriou, F.: Numerical models for rotor cage induction machines using finite element method. IEEE Trans. Magn. **34**(5), 3202–3205 (1998)
3. Rapetti, F.: An overlapping mortar element approach to coupled magneto-mechanical problems. Math. Comput. Simul. **88**(8), 1647–1656 (2010)

4. Liu, Z., Liu, S., Mohammed, O.A.: A practical method for building the fe-based phase variable model of single phase transformers for dynamic simulations. IEEE Trans. Magn. **43**(4), 1761–1764 (2007)
5. Lee, S.M., Lee, S.H., Choi, H.S., Park, I.H.: Reduced modeling of eddy current-driven electromechanical system using conductor segmentation and circuit parameters extracted by FEA. IEEE Trans. Magn. **41**(5), 1448–1451 (2005)
6. Schilders, W.H., Van der Vorst, H.A., Rommes, J.: Model Order Reduction: Theory, Research Aspects and Applications. Springer, Berlin (2008)
7. Yue, Y., Feng, L., Meuris, P., Schoenmaker, W., Benner, P.: Application of Krylov-type parametric model order reduction in efficient uncertainty quantification of electro-thermal circuit models. In: Proceedings of the Progress in Electromagnetics Research Symposium (PIERS), pp. 379–384 (2015)
8. Banagaaya, N., Feng, L., Meuris, P., Schoenmaker, W., Benner, P.: Model order reduction of an electro-thermal package model. IFAC-PapersOnLine **48**(1), 934–935 (2015)
9. Albunni, M.N., Rischmuller, V., Fritzsche, T., Lohmann, B.: Model-order reduction of moving nonlinear electromagnetic devices. IEEE Trans. Magn. **44**(7), 1822–1829 (2008)
10. Henneron, T., Clénet, S.: Model order reduction applied to the numerical study of electrical motor based on POD method taking into account rotation movement. Int. J. Numer. Model. **27**(3), 485–494 (2014)
11. Sato, T., Sato, Y., Igarashi, H.: Model order reduction for moving objects: fast simulation of vibration energy harvesters. COMPEL **34**(5), 1623–1636 (2015)
12. Schmidthäusler, D., Schöps and Clemens, M.: Linear subspace reduction for quasistatic field simulations to accelerate repeated computations. IEEE Trans. Magn. **50**(2), article # 7010304 (2014)
13. Karl, H., Fetzer, J., Kurz, S., Lehner, G., Rucker, W.M.: Description of TEAM workshop problem 28: an electrodynamic levitation device. In: Proceedings of the TEAM Workshop, Graz, pp. 48–51 (1997)
14. Henrotte, F., Nicolet, A., Hedia, H., Genon, A., Legros, W.: Modelling of electromechanical relays taking into account movement and electric circuits. IEEE Trans. Magn. **30**(5), 3236–3239 (1994)
15. Sato, Y., Igarashi, H.: Model reduction of three-dimensional eddy current problems based on the method of snapshots. IEEE Trans. Magn. **49**(5), 1697–1700 (2013)
16. Hasan, Md. R., Sabariego, R. V., Geuzaine, C., Paquay, Y.: Proper orthogonal decomposition versus Krylov subspace methods in reduced-order energy-converter models. In: Proceedings of IEEE International Energy Conference, pp. 1–6 (2016)
17. Parent, G., Dular, P., Ducreux, J.P., Piriou, F.: Using a Galerkin projection method for coupled problems. IEEE Trans. Magn. **44**(6), 830–833 (2008)

Chapter 21
Time-Domain Reduced-Order Modelling of Linear Finite-Element Eddy-Current Problems via RL-Ladder Circuits

Ruth V. Sabariego and Johan Gyselinck

Abstract This paper deals with the reduced-order modelling of magnetically-linear eddy-current devices which have a single electrical port, i.e. one terminal voltage and current. The device is first characterised by means of frequency-domain finite-element computations considering the relevant frequency interval, for subsequently fitting constant-coefficient RL ladder circuits of adjustable size (i.e. number of branches and loops). The accuracy of the ladder-circuit model is assessed in both frequency and time domain. This approach is successfully applied to the axisymmetric magnetic-levitation device of TEAM Workshop problem 28, which includes a position degree of freedom as well.

21.1 Introduction

Finite-element (FE) modelling of electromagnetic devices allows to consider with variable precision the (2D or 3D) geometry and dimensions, magnetic material properties (including saturation and possibly hysteresis), time variation (static, sinusoidal regime and time stepping), induced currents in massive conducting parts, electrical connectivity and supply conditions, along with mechanical conditions and equations in case of an electrical machine or actuator [1, 2]. In high-frequency devices and high-speed machines, eddy currents and eddy-current effects in lamination stacks and windings can be of prime concern, in which case homogenization techniques, rather than brute-force modelling, may be indispensable for computational-cost reasons [3, 4].

R. V. Sabariego (✉)
KU Leuven, Dept. Electrical Engineering, Campus EnergyVille, Genk, Belgium
e-mail: ruth.sabariego@kuleuven.be

J. Gyselinck
Université Libre de Bruxelles, BEAMS Dept., Brussels, Belgium
e-mail: johan.gyselinck@ulb.ac.be

© Springer International Publishing AG, part of Springer Nature 2018
U. Langer et al. (eds.), *Scientific Computing in Electrical Engineering*,
Mathematics in Industry 28, https://doi.org/10.1007/978-3-319-75538-0_21

Time-stepping FE simulation may be the suitable approach for studying the performance of the device in simple steady-state electrical and mechanical operating conditions, and possibly for optimising the device in such conditions. If the next stage in the analysis/design is about more realistic transient operation, possibly along with (closed-loop) control and integration in the wider system, the FE model may become prohibitively expensive, and a computationally cheaper model, generically referred to as reduced-order model (ROM), should be derived, even when this implies a certain loss of accuracy.

Purely mathematical ROMs [5] in various engineering disciplines are flexible and automated models that capture the essential features of a problem with a prescribed accuracy. The most widely used ROMs are those based on the proper orthogonal decomposition (POD) [6]. The POD approach relies on a snapshot selection, i.e. solutions of the full model for typical working conditions, at different frequencies or time steps, so as to construct a reduced basis [7]. Expensive Greedy algorithms that scan the discrete space are used for selecting suitable sets of snapshots [8]. In [9], POD is combined with discrete empirical interpolation for handling a nonlinear magnetostatic problem. The treatment of movement has been considered as well, though there are to date very few works in electromagnetics. An electrical machine is studied in [10] with a locked-step approach, i.e. the mesh does not change with rotation. As common feature, all these ROMs involve little or no physical insight of the device or problem at hand.

Alternatively, ROMs can also be based on the systematic, more physics-based identification of the device via a series of simple static or dynamic FE computations [11, 12], rather than on a manipulation of the FE equations and matrices. Such an approach is more or less straightforward and feasible depending on the characteristics of the electromagnetic device: the number of independent currents (or voltages) n_i, the absence/presence of saturable magnetic materials and induced currents, and the number of position degrees of freedom n_p (none, one or more). Different particular cases can be distinguished. If the system is magnetically linear and comprises no eddy currents, the current-independent $n_i \times n_i$ inductance matrix can be straightforwardly obtained by n_i magnetostatic computations; position dependence, if any, can be quantified by a suitable sampling in the n_p-dimensional motion space. Through tabulation and interpolation (extrapolation) of the inductance values, a minimum-cost model of the device is easily arrived at, with a priori little or no loss of accuracy.

The situation changes drastically in presence of saturable material, as all $n_i \times n_i$ inductance values depend on all n_i currents, leading quickly (for all but very small n_i) to an unworkable number of magnetostatic computations to carry out for sampling the current space, possibly combined with the motion space [13]. The presence of eddy currents (or eddy-current effects) is another major complication. For linear (i.e. magnetically-unsaturable) systems, frequency-domain characterisation is relatively straightforward, as well as its time-domain extension. The latter can be easily done via a ladder-circuit-like approach for single-port systems. For a generalisation to multi-port systems, see e.g. [14]. For the case with eddy currents and saturation, a general approach is much less evident and ad-hoc approaches and approximations seem unavoidable, e.g. [15].

In this paper we focus on a relatively simple, though far from trivial case, which is the one of a magnetically-linear device having one terminal current (and voltage) and which includes eddy currents and 1D movement. The developments are directly done for and applied to the levitation device of TEAM Workshop problem 28 (TWP28) [16], knowing that the application to 2D translation-symmetrical and 3D problems is very similar. In the following section the TWP28 device is first presented and identified in the frequency domain, at a number of relevant frequencies and positions. In Sect. 21.3 an approximate ladder-circuit model is fitted and validated considering both sinusoidal and transient regime, whereby the position is held constant (within the relevant range). Note that time-varying position is a non-trivial complication, which will be dealt with in a further paper.

21.2   TWP28: Frequency-Domain FE Identification

The axisymmetric device of TWP28 comprises two concentric anti-series-connected coils (both of rectangular cross-section in the rz-plane) and a 3 mm-thick aluminum circular disk located concentrically at a certain height above the coils [16]. Part of the 2D model can be seen in Figs. 21.1 and 21.2; the clearance between the coils and the plate is 3 mm, i.e. position degree of freedom z_{pl}.

The well-known magnetodynamic vector potential formulation (MVP), in terms of its tangential component is adopted, either in the time or frequency domain ($a_\phi(t)$ or $\underline{a}_\phi$); its allows for tangential current density, which is either imposed

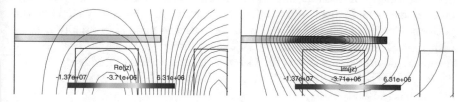

Fig. 21.1 Flux lines and induced current density in the plate, real (left) and imaginary (right) parts, with $f = 50\,\mathrm{Hz}$, $z_{pl} = 3\,\mathrm{mm}$ and $\underline{i} = 20\,\mathrm{A}$

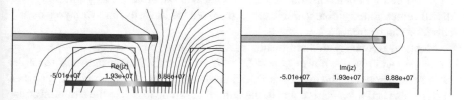

Fig. 21.2 Flux lines and induced current density in the plate, real (left) and imaginary (right) parts, with $f = 1\,\mathrm{kHz}$, $z_{pl} = 3\,\mathrm{mm}$ and $\underline{i} = 20\,\mathrm{A}$

or induced (stranded versus massive conductors, terminology used in [17]). All complex numbers are denoted by underlined symbols. E.g., a sinusoidal current $i(t) = \hat{i}\cos(\omega t + \gamma)$ is represented by phasor $\underline{i} = \hat{i}\exp(\underline{j}\gamma) = \hat{i}(\cos\gamma + \underline{j}\sin\gamma)$, where f and $\omega = 2\pi f$ are the frequency and the pulsation, $\hat{i}$ the amplitude, γ the phase angle, and $\underline{j}$ the imaginary unit.

The inner and outer coil have $n_1 = 960$ and $n_2 = 576$ turns of copper wire respectively, for a total DC resistance of $R_0 = 6.73\,\Omega$; no skin and proximity effect in the winding is considered. The classical, trivial stranded-coil modelling implies a uniform current density in the two respective cross-sections [4], and allows for either imposed terminal current, $i(t)$ or $\underline{i}$, imposed terminal voltage, $v(t)$ or $\underline{v}$, or insertion in an electrical current [17]. The associated flux-linkage of the anti-series connected coils, $\Psi(t)$ or $\underline{\Psi}$, simply depends on the MVP, $a_\phi(r, z, t)$ or $\underline{a}_\phi(r, z)$, via the integral of the latter over the cross-section of the coils, $\Omega_{\text{coil},1}$ and $\Omega_{\text{coil},2}$, considering the respective number of turns [17]:

$$\psi(t) = \frac{n_1\,2\pi}{\Omega_{\text{coil},1}}\int_{\Omega_{\text{coil},1}} a_\phi\,r\mathrm{d}r\mathrm{d}z - \frac{n_2\,2\pi}{\Omega_{\text{coil},2}}\int_{\Omega_{\text{coil},2}} a_\phi\,r\mathrm{d}r\mathrm{d}z. \tag{21.1}$$

The three global variables, i.e. voltage, current and flux-linkage, are linked, either instantaneously or phasor-wise, as follows:

$$v(t) = R_0\,i(t) + \frac{\mathrm{d}\psi}{\mathrm{d}t} \quad \text{and} \quad \underline{v} = R_0\,\underline{i} + \underline{j}\omega\underline{\Psi}. \tag{21.2}$$

In the low-frequency limit $f \to 0$, there are no eddy currents in the (non-magnetic) plate, such that $\psi = L_0 i$ with $L_0 = 73.23\,\mathrm{mH}$ the DC inductance. This inductance is in theory z_{pl}-independent as the plate is non-magnetic. In practice, using a position-dependent FE mesh, with around 4600 triangular first-order elements and around 2000 nodes, the resulting L_0 varies very slightly with position, namely less than $\pm 0.5\,\mu\mathrm{H}$ or $\pm L_0/10^5$. The plate is invariably meshed with four layers of elements in the thickness direction, which is sufficient for frequency-domain computations till 1 kHz; indeed, at the latter frequency the skin depth $\sqrt{2/(\omega\sigma\mu_0)}$ in the plate is 2.7 mm, such that the 3-layer-per-skin-depth rule of thumb is satisfied.

The complex terminal impedance $\underline{Z} = \underline{v}/\underline{i} = R + \underline{j}\omega L$ depends on both frequency f and position z_{pl}. The series AC resistance $R(f, z_{\text{pl}})$ and inductance $L(f, z_{\text{pl}})$ can be obtained from the FE model, with suitable $\{f, z_{\text{pl}}\}$ sampling, via the aforementioned global variables, where, given the linearity, any voltage $\underline{v}$ or any current $\underline{i}$ can be imposed.

Identical frequency-domain results can be obtained via the active power $P = \frac{1}{2}R\,\hat{i}^2$ (W) and reactive power $Q = \frac{1}{2}L\,\hat{i}^2$ (W, or rather VAr, volt-amperes-reactive, as is conventional in the electrotechnical community), with e.g. imposed $\underline{i} = 1\,\mathrm{A}$ current. The active power is equal to the sum of the Joule losses in the stranded coils, $\frac{1}{2}R_0\,\hat{i}^2$, and those in the conducting domain; the latter follow from the integration of the loss density $\frac{1}{2}\rho\hat{j}_\phi^2$ over the disk cross-section Ω_{disk}, with ρ the resistivity and $\hat{j}_\phi(r,z)$ the amplitude of the current density:

$$P = \frac{1}{2}R\hat{i}^2 = \frac{1}{2}R_0\,\hat{i}^2 + \frac{1}{2}\int_{\Omega_{\text{disk}}} \rho\hat{j}_\phi^2\,2\pi r\mathrm{d}r\mathrm{d}z\,. \tag{21.3}$$

The reactive power Q follows from the integration of the magnetic energy density $\frac{1}{2}\nu\hat{b}^2$, with ν the reluctivity and $\hat{b}^2(r,z) = \hat{b}_r^2 + \hat{b}_z^2$ the square norm of the flux density vector, over the complete cross-section Ω of the model:

$$Q = \frac{1}{2}L\hat{i}^2 = \frac{1}{2\omega}\int_{\Omega} \nu\hat{b}^2\,2\pi r\mathrm{d}r\mathrm{d}z\,. \tag{21.4}$$

Some results obtained this way are shown in Fig. 21.3, namely the relative change in resistance and inductance due to the induced current in the conducting disk, i.e. $\Delta R(f, z_{\text{pl}})/R_0$ and $-\Delta L(f, z_{\text{pl}})/L_0$, with $\Delta R = R - R_0$ and $\Delta L = L - L_0$, versus frequency, in the 10–1000 Hz interval, with a suitable double logarithmic scale, and for three separate positions, viz $z_{\text{pl}} = 3$, 10 and 17 mm.

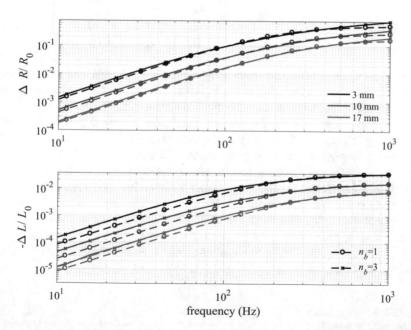

Fig. 21.3 Relative increase/decrease of terminal AC resistance and inductance versus frequency, for three different positions, obtained with the FE model (full lines) and the ladder circuit (markers, $n_b = 1$ and $n_b = 3$)

21.3   Ladder-Circuit Approximation

The frequency-dependent series resistance and inductance, $R(f, z_{pl})$ and $L(f, z_{pl})$, obtained with the FE model and depicted in Fig. 21.3 (full lines), can be approximately effected with a finite constant-coefficient ladder circuit as shown in Fig. 21.4. Fed by the terminal voltage $v(t)$ (terminal current $i(t)$), this circuit comprises a loop with the DC resistance R_0 and DC inductance L_0 and further n_b auxiliary loops $\{R_k, L_k\}$ and currents i_k $(k = 1 \dots n_b)$. The $n_b + 1$ circuit equations can be written in matrix notation in terms of the column vector $\mathbf{I}(t) = [i(t)\, i_1(t)\, i_2(t)\, \dots\, i_{n_b}(t)]^T$ or $\underline{\mathbf{I}}$ and corresponding voltage column vector $\mathbf{V}(t) = [v(t)\, 0\, 0\, \dots\, 0]^T$ or $\underline{\mathbf{V}}$:

$$\mathbf{V}(t) = \mathbf{R}\,\mathbf{I}(t) + \mathbf{L}\frac{d}{dt}\mathbf{I}(t) \quad \text{or} \quad \underline{\mathbf{V}} = \left(\mathbf{R} + j\omega\mathbf{L}\right)\underline{\mathbf{I}}, \tag{21.5}$$

where $\mathbf{R}$ is diagonal and $\mathbf{L}$ tridiagonal and symmetric. Note that the source term in (21.5) is either the terminal current i or the terminal voltage v, i.e. the first element in column vectors $\mathbf{V}(t)$ or $\mathbf{I}(t)$ ($\underline{\mathbf{I}}$ or $\underline{\mathbf{V}}$), respectively.

With $n_b = 2$, i.e. the ladder circuit shown in Fig. 21.4, $\mathbf{R}$ and $\mathbf{L}$ read:

$$\mathbf{R} = \text{diag}\,(R_0, R_1, R_2)\,, \qquad \mathbf{L} = \begin{bmatrix} L_0 & -L_0 & 0 \\ -L_0 & L_0 + L_1 & -L_1 \\ 0 & -L_1 & L_1 + L_2 \end{bmatrix}. \tag{21.6}$$

In near-DC conditions, low-frequency limit $f \to 0$, this circuit amounts to the terminal DC series impedance $R_0 + j\omega L_0$. As a reminder, for the application considered, TWP28, this impedance is position-independent.

For a given n_b and z_{pl}, the $2n_b$ parameters R_k and L_k are determined by fitting the ensuing impedance $\underline{Z}_{n_b}(f, z_{pl})$ to the reference FE impedance $\underline{Z}_{FE}(f, z_{pl})$ in the relevant frequency interval, e.g. by means of the Nelder-Mead simplex method (nonlinear minimization). Some results are depicted in Fig. 21.3 for $n_b = 1$ and $n_b = 3$; one observes an excellent agreement with the FE results for $n_b = 3$.

Next time-domain computations with $f = 1\,\text{kHz}$, $v(t) = 500\,\text{V} \cdot \sin(\omega t)$ ($\underline{v} = -j\,500\,\text{V}$) and z_{pl} equal to either 3, 10 or 17 mm are carried out. With the ladder-circuit approximation, the instantaneous Joule losses in the plate are given by $\sum_{k=1}^{n_b} R_k\, i_k^2(t)$. See Fig. 21.5. Excellent convergence towards the FE results is observed with increasing n_b.

Fig. 21.4 Ladder circuit with two auxiliary loops and currents ($n_b = 2$)

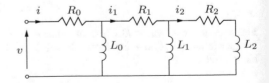

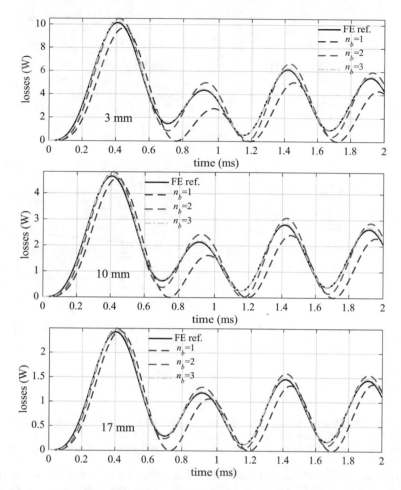

Fig. 21.5 Losses in the plate versus time ($z_{pl} = 10$ mm; 1 kHz voltage supply), computed with FE model and RL ladder circuit (n_b equal to 1, 2 and 3)

Figure 21.6 shows how the resistance and inductance values of the ladder circuit with $n_b = 3$ vary with the position z_{pl} of the plate, as a result of the separate fitting for each discrete position. The six curves shown are each smooth, but some curves comprise a deflection point.

Further development, in a next paper, will include time-varying position. For this, one single global fitting may be required, using for the resistances and the inductances preset expressions in terms of z_{pl}.

Acknowledgements Work supported in part by the Walloon Region of Belgium (WBGreen FEDO, grant RW-1217703) and the Belgian Science Policy (IAP P7/02).

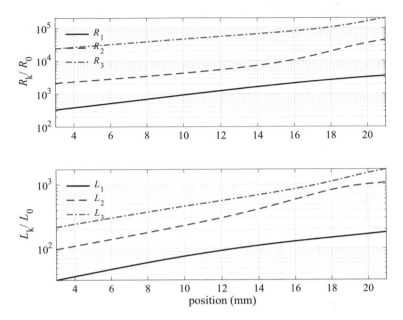

Fig. 21.6 Fitted R_k resistances and L_k inductances of the $n_b = 3$ ladder circuit versus position (fitting between 10 Hz and 1 kHz, per position, with $\Delta z_{pl} = 0.25$ mm)

References

1. Schmidt, E.: Finite element analysis of electrical machines and transformers. COMPEL - Int. J. Comput. Math. Electr. Electron. Eng. **30**(6), 1899–1913, (2011)
2. Meunier, G.: The Finite Element Method for Electromagnetic Modeling. Wiley, New York (2008)
3. Gyselinck, J., Sabariego, R.V., Dular P.: A nonlinear time-domain homogenization technique for laminated iron cores in three-dimensional finite-element models. IEEE Trans. Magn. **42**(4), 763–766 (2006)
4. Gyselinck, J., Sabariego, R.V., Dular P.: Time-domain homogenization of windings in 2-D finite element models. IEEE Trans. Magn. **43**(4), 1297–1300 (2007)
5. Schilders, W., der Vorst, H. V., Rommes, J.: Model Order Reduction: Theory, Research Aspects and Applications. Springer, New York (2008)
6. Sato, Y., Igarashi, H.: Model reduction of three-dimensional eddy current problems based on the method of snapshots. IEEE Trans. Magn. **49**(5), 1697–1700 (2013)
7. Paquay, Y., Geuzaine, C., Hasan, Md.R., Sabariego, R.V.: Reduced order model for accounting for high frequency effects in power electronic components. IEEE Trans. Magn. **52**(3), article #: 7202904 (2016)
8. Buffa, A., Maday, Y., Patera, A.T., Prudhomme, C., Turinici, G.: A priori convergence of the Greedy algorithm for the parametrized reduced basis method. ESAIM: Math. Modell. Numer. Anal. **46**(3), 595–603 (2012)
9. Henneron, T., Clénet, S.: Model order reduction applied to the numerical study of electrical motor based on POD method taking into account rotation movement. Int. J. Numer. Model. **27**(3), 485–494, (2014)
10. Henneron, T., Clénet, S.: Model order reduction of non-linear magnetostatic problems based on POD and DEI methods. IEEE Trans. Magn. **50**(2), 33–36 (2014)

11. Lee, S.-M., Lee, S.-H., Choi, H.-S., Park, I.-H.: Reduced modeling of eddy current-driven electromechanical system using conductor segmentation and circuit parameters extracted by FEA. IEEE Trans. Magn. **41**(5), 1448–1451 (2005)
12. Liu, Z., Liu, S., Mohammed, O.A.: A practical method for building the FE-based phase variable model of single phase transformers for dynamic simulations. IEEE Trans. Magn. **43**(4), 1761–1764 (2007)
13. Rasilo, P., Lemesle, M., Belahcen, A., Arkkio, A., Hinkkanen, M.: Comparison of finite-element-based state-space models for PM synchronous machines. IEEE Trans. Energy Convers. **20**(2), 535–543 (2014)
14. De Greve, Z., Deblecker, O., Lobry, J., Kéradec, J. P.: High-frequency multi-winding magnetic components: from numerical simulation to equivalent circuits with frequency-independent RL parameters. IEEE Trans. Magn. **50**(2), 141–144 (2014)
15. Shindo, Y., Miyazaki, T., Matsuo, T.: Cauer circuit representation of the homogenized eddy-current field based on the Legendre expansion for a magnetic sheet. IEEE Trans. Magn. **52**(3), 1–4 (2016)
16. Hollaus, K., Fetzer, J., Kurz, S., Lehner, G., Rucker, W. M.: Description of TEAM workshop problem 28: an electrodynamic levitation device. In: Proceedings of the TEAM Workshop, Graz, pp. 41–42 (1997)
17. Lombard, P., Meunier. G.: A general purpose method for electric and magnetic combined problems for 2D, axisymmetric and transient systems. IEEE Trans. Magn. **29**(2), 1737–1740 (1993)

Part VI
Industrial Applications

The last part of this book comprises two contributions concerning *Industrial Applications*. There were several presentations on different aspects of industrial applications at the Industrial Day, and at the first poster session, which was also devoted to applications.

The paper by M. Cremonesi et al. on *"A Lagrangian Approach to the Simulation of a Constricted Vacuum Arc in a Magnetic Field"* was presented by Massimiliano Cremonesi at the invited session of Industrial Day. The paper proposes to use the Lagrangian description of a minimal arc model in 2d as starting point for the discretization. The authors present selected numerical results obtained by means of their approach, and compare these results with the simulation results computed by the commercial package ANSYS Fluent.

ABB has developed a simulation framework for the dielectric design of high-voltage devices called the *Virtual High Voltage Lab (VHVLab)*. This simulation framework was presented by Andreas Blaszczyk in the first poster session on industrial applications. The corresponding paper by A. Blaszczyk et al. provides a description of the basic concept and the components of a VHVLab. Furthermore, the authors demonstrate the use of the VHVLab in real applications from ABB.

Chapter 22
A Lagrangian Approach to the Simulation of a Constricted Vacuum Arc in a Magnetic Field

Massimiliano Cremonesi, Attilio Frangi, Kai Hencken, Marcelo Buffoni, Markus Abplanalp, and Jörg Ostrowski

Abstract The use of numerical simulations of vacuum arcs can be very useful in order to improve the performance of vacuum interrupters. Standard computational fluid dynamics methods based on the Eulerian approach have difficulties to deal with this kind of problem, so a new technique is proposed, based on a Lagrangian approach. In order to focus on the performance of the new approach and not on specific details of a full model, a simplified arc model is used to investigate the capabilities of a Lagrangian approach in the context of vacuum arc simulations. The focus of this initial study is on implementing the necessary ingredients, that is, the development of a compressible flow solver, the introduction of the relevant boundary conditions and the coupling with the current conservation equation for the electric current. In addition, the stability of such a numerical scheme is evaluated. Furthermore, comparisons with results obtained using commercial software are also provided to demonstrate the validity of the results obtained with the new methodology.

22.1 Introduction

Vacuum interrupters are electrical protection devices that are used extensively in the medium voltage range up to 40 kV in order to both switch nominal currents of a few kA and interrupt fault currents up to 63 kA. They rely on the insulation properties of the vacuum in the gap between two electrodes for their voltage withstand capability. When interrupting current, a vacuum arc is formed for a time of up to a half-wave.

M. Cremonesi (✉) · A. Frangi
Politecnico di Milano, Milano, Italy
e-mail: massimiliano.cremonesi@polimi.it; attilio.frangi@polimi.it

K. Hencken · M. Buffoni · M. Abplanalp · J. Ostrowski
ABB Switzerland Ltd., Corporate Research, Baden, Switzerland
e-mail: kai.hencken@ch.abb.com; marcelo.buffoni@ch.abb.com; markus.abplanalp@ch.abb.com; joerg.ostrowski@ch.abb.com

© Springer International Publishing AG, part of Springer Nature 2018 243
U. Langer et al. (eds.), *Scientific Computing in Electrical Engineering*,
Mathematics in Industry 28, https://doi.org/10.1007/978-3-319-75538-0_22

A vacuum arc is a metal-vapor thermal arc formed by the metallic material evaporated from the two electrodes and heated to a high temperature by the ohmic heating. This type of plasma arc shows some different modes depending on current and gap distance, see e.g., Chapter 2 of [1]. Under certain conditions, the arc is in a constricted, columnar mode which means that it does not cover the full contacts (anode and/or cathode) areas and is in a high density and pressure state [2, 3]. The energy given to the cathode and anode through the exchange of electrons and ions heats them up to such high temperatures that the arc roots are able to provide new metal vapor for the plasma arc to sustain itself. Interruption of large currents can only occur at a natural zero crossing of the current. At that moment, the current is low, the energy input is small, and the gap can be cleared of the plasma and metal vapor. However, successful current interruption can occur only if the metal vapor density in the gap between the two contacts and the surface temperature of both electrodes, are below a critical value.

To achieve the required conditions for successful interruption, some arc control is needed. Two main arc control principles have evolved that try to spread the heating of the contacts through the arc over a larger area to reduce the temperature, and therefore the evaporation at current zero. The so-called AMF approach tries to keep the arc in the non-constricted, diffuse state with the help of an axial magnetic field called the AMF approach. The other commonly used approach is to use the so-called "transverse magnetic field" (TMF) principle. In this approach, the focus of this work, electrode shapes are used to generate a magnetic field which is predominantly transverse to the gap direction and the arc. The magnetic field generates a Lorentz force that moves the arc with a high velocity over the surface of the electrodes, spreading the heat flux over a larger area, thereby reducing the maximal surface temperature.

In order to improve the performance of such vacuum interrupters, it is fundamental to understand the motion of the arc. Since experimental investigations are difficult due to the extreme conditions and the fast motion of the arc, numerical simulations are preferable. A detailed arc model was developed in [4–7] and its capability was demonstrated, in principle, to simulate the movement of an arc. In order to be useful for the development of new interrupters, the simulations need to be fast and reliable. This is difficult for two reasons. First, there are strong gradients at the edge of the vacuum arc where the pressure decreases from several bars in the core of the arc to values well below 1 millibar over a very short distance. As the temperature does not change greatly, the density follows a similar steep decrease. Standard computational fluid dynamics numerical formulations, based on the Eulerian approach, have severe difficulties in dealing with such strong gradients. Second, the TMF arc typically fills only a small part of the simulation domain. This means that a large portion of the mesh consists of cells having vacuum conditions, that is, where the solution is neither useful nor relevant. Although using an adaptive mesh is one possible option to cope with this issue, we are interested in exploring a different and innovative approach which automatically generates a fine mesh in the arc region. In the extreme case, the Lagrangian approach described in the following sections enables the definition of a mesh only in those regions where the arc plasma is present. It is therefore potentially very attractive for such simulations.

In the last decade, Lagrangian approaches have received considerable attention, in particular for incompressible flow simulations with strong free surface evolution [10]. Indeed, while in the Eulerian framework the problem is solved on a fixed mesh through which the fluid moves, in Lagrangian methods the mesh follows the fluid which makes tracking the free surface easier. Applications to casting of complex fluids [8] or landslide simulation [9] are only examples of the possible domains of interest and testify to the robustness of the technique. Over the past few years, the analysis of compressible flows has emerged as a promising field for application of Lagrangian methods [11, 12], in particular in those contexts where standard techniques lose robustness and are therefore less attractive to be used. Due to the presence of shocks, viscous boundary layers, mixing of different constituents and strong density gradients, the numerical solution of the arc equations can pose severe challenges.

The main aim of this first investigation is therefore *in primis* to provide a preliminary demonstration of the applicability of the Lagrangian approach to arc simulation. Therefore we focus on specific issues, like the use of a compressible flow solver, the introduction of evaporating and absorbing boundary conditions and the stability of the approach under the difficult physical conditions. In order to assess the validity of the new numerical approach a comparison is made with a state-of-the-art CFD calculation.

Clearly this only represents a preliminary step towards the full arc simulation, which requires the movement of the plasma arc in a strongly coupled multi-physics problem to be tracked.

22.2   Minimal Arc Model

In order to perform a preliminary investigation of the capability of the Lagrangian approach to address vacuum arc simulations, we define here a "minimal arc model" in a 2D context. Implementing the full model is described in [4–7] and should be the ultimate aim of the simulation. The minimal model should be at the same time as close as possible to the real vacuum arc, capturing possible dependencies of fluid parameters, and be based on equations that can be easily implemented. Defining ρ the density, $\mathbf{u} = u_x \mathbf{e}_x + u_y \mathbf{e}_y$ the velocity field, e the internal energy, the balance equations for a compressible fluid subjected to a prescribed magnetic field $\mathbf{B} = B_x \mathbf{e}_x$ can be written as:

$$\text{mass conservation:} \quad \frac{1}{\rho}\frac{d\rho}{dt} = -\nabla \cdot \mathbf{u} \tag{22.1}$$

$$\text{momentum conservation:} \quad \rho\frac{d\mathbf{u}}{dt} = -\nabla p + \nabla \cdot \boldsymbol{\tau} + \mathbf{J} \times \mathbf{B} \tag{22.2}$$

energy conservation: $\rho \dfrac{de}{dt} = -p\,\nabla \cdot \mathbf{u} + \boldsymbol{\tau} : \nabla \mathbf{u} + \nabla \cdot (k_e \nabla T) + \dfrac{\mathbf{J}^2}{\sigma} + Q_{rad}$

$$\text{(22.3)}$$

current conservation: $\nabla \cdot (\sigma \nabla \phi) = 0$ (22.4)

where p is the pressure field and $\boldsymbol{\tau}$ the deviatoric stress tensor defined as:

$$\boldsymbol{\tau} = \mu \left[\nabla \mathbf{u} + \nabla \mathbf{u}^T - \frac{2}{3} (\nabla \cdot \mathbf{u}) \mathbf{I} \right] \qquad \text{(22.5)}$$

where $\mathbf{J}$ the current flux and ϕ the potential are related by:

$$\mathbf{J} = -\sigma \nabla \phi \qquad \text{(22.6)}$$

It should be stressed that in the first three equations the total (material) derivative on the left hand side includes the transport terms which are by contrast explicitly expressed in standard Eulerian approaches. Q_{rad} is a loss term due to radiation. In principle, the radiation losses have to be calculated by a separate transport equation. Instead a "net-emission" approach, that only describes the loss of power as a function of density and temperature, is often used. Here, we simplify this even further and define $Q_{rad} = -\sigma_{sb} T^4 / R$, where σ_{sb} is the Stefan-Boltzmann constant and R a parameter (in the following $R = 0.01\,\text{m}$). By appropriately choosing R, the arc temperature can be tuned to a realistic value. For the plasma properties we use an ideal gas equation of state with adiabatic coefficient, heat capacity and internal energy extracted in tabular form from more detailed calculations. Transport properties have been calculated by using the Spitzer equations for partially ionized plasmas [13]. Only Cu and Cu^+ were considered with their densities calculated from the Saha equation. This has the advantage that all equations can be solved explicitly as a function of pressure and temperature and still be rather realistic.

As mentioned in the introduction, the magnetic field plays an important role in defining the arc. In principle, it needs to be calculated self-consistently from the current flow. As we want to focus on the plasma flow, we have chosen to use an assigned magnetic field. In order to mimic the pinching force of a real field, we define B_x as

$$B_x = \begin{cases} \mu_0 \bar{J} \min(x, d) & \text{if } x > 0 \\ \mu_0 \bar{J} \max(x, -d) & \text{if } x < 0 \end{cases} \qquad \text{(22.7)}$$

where μ_0 is the permeability, x the distance from the center of the arc root and d the diameter; in our case $\bar{J} = 3 \times 10^8\,\text{A/m}^2$ and $d = 0.5\,\text{cm}$.

22.3 Numerical Method

The conservation equations have been solved using a Lagrangian finite element approach. A complete discussion and validation of the formulation for compressible flow problems can be found in the recently published research paper [14]; only some key elements are presented here for brevity. Starting from the weak form of the equations, a standard finite element space discretization has been applied, leading to the following semi-discretized equations:

$$\mathbf{M}_\rho(t)\mathbf{R}(t) = \mathbf{R}_0 \tag{22.8}$$

$$\mathbf{M}_u \frac{d\mathbf{U}}{dt} = \mathbf{F}_u(\mathbf{U}(t), \mathbf{E}(t), \mathbf{R}(t)) \tag{22.9}$$

$$\mathbf{M}_e \frac{d\mathbf{E}}{dt} = \mathbf{F}_e(\mathbf{U}(t), \mathbf{E}(t), \mathbf{R}(t)) \tag{22.10}$$

$$\mathbf{L}\boldsymbol{\phi} = \mathbf{0} \tag{22.11}$$

where $\mathbf{M}_u$ is the "velocity" mass matrix, $\mathbf{M}_e$ the "energy" mass matrix, and $\mathbf{M}_\rho$ "density' mass matrix. $\mathbf{F}_u$ and $\mathbf{F}_e$ represent the internal forces for momentum and energy conservation equations, respectively. For the time integration of mass, momentum and energy equations, a forward Euler scheme has been used, leading to an explicit solution scheme. Current conservation is, in contrast, intrinsically implicit. The fully discretized system reads:

$$\mathbf{M}_\rho^n \mathbf{R}^n = \mathbf{R}^0 \tag{22.12}$$

$$\mathbf{M}_u \mathbf{U}^{n+1} = \mathbf{M}_u \mathbf{U}^n + \Delta t \mathbf{F}_u(\mathbf{U}^n, \mathbf{E}^n, \mathbf{R}^n, \mathbf{P}^n) \tag{22.13}$$

$$\mathbf{M}_e \mathbf{E}^{n+1} = \mathbf{M}_e \mathbf{E}^n + \Delta t \mathbf{F}_e(\mathbf{U}^{n+1/2}, \mathbf{E}^n, \mathbf{R}^n, \mathbf{P}^n) \tag{22.14}$$

$$\mathbf{L}\boldsymbol{\phi}^{n+1} = \mathbf{0} \tag{22.15}$$

where $\mathbf{U}^{n+1/2} = \frac{1}{2}(\mathbf{U}^{n+1} + \mathbf{U}^n)$ is used to preserve total energy from time t^n to time t^{n+1} [12, 15]. Matrices $\mathbf{M}_u$ and $\mathbf{M}_e$ are constant as long as the reference configuration is fixed. Moreover, mass lumping is introduced and the velocity and energy vectors can thus be evaluated node by node without requiring the solution of linear systems. Finally, following the Lagrangian nature of the proposed approach, the nodal coordinates should be updated:

$$\mathbf{x}^{n+1} = \mathbf{x}^n + \Delta t \mathbf{U}^{n+1} \tag{22.16}$$

Equation (22.16) imposes that at every time step the position of the mesh nodes moves following the fluid velocity, leading to possible mesh distortion. In the solution scheme proposed, a check of the mesh is performed at every time step. When the mesh is too distorted, a Delaunay tessellation is used to generate a new regular mesh. Starting from the current position of the points, the Delaunay tessellation creates a new connectivity using only nodal values for the unknowns, so that no mapping is necessary from the old mesh to the new one.

22.3.1 Boundary Conditions

In the specific application at hand, both inflow and outflow boundary conditions are present. In the former case, mass, momentum and energy flux are imposed (see next section). In a Lagrangian framework this can be easily achieved by enforcing the desired values of density, velocity and energy to nodes on the in-flow boundary portion; these nodes eventually move away but are continuously replaced with new particles. In the latter case, outflow is simulated by enforcing zero material derivative of density, velocity and energy on particles which exit the simulation domain; these enter an artificial "buffer" and are eventually eliminated.

22.4 Simplified Arc Test and Numerical Results

We have also defined a simplified 2D test geometry, which should mimic the real one (see Fig. 22.1).

A plasma simulation requires thermodynamic and transport properties and in addition suitable boundary conditions as input. All boundaries are defined to be perfectly transparent (outflow) apart from the region that defines the arc roots. Here the evaporation is written in the form of a Hertz-Knudsen evaporation model with

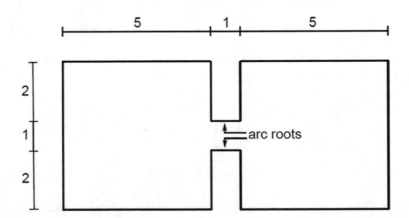

Fig. 22.1 Geometry of the test case. Measures are expressed in cm

the mass flux given by

$$\frac{m_{Cu}}{\sqrt{2\pi m_{Cu}kT_s}}(p^* - p)$$

as well as the momentum flux

$$\frac{1}{2}(p + p^*)$$

and the energy flux

$$\frac{2kT_s}{\sqrt{2\pi m_{Cu}kT_s}}(p^* - p)$$

where m_{Cu} is the mass of an individual copper atom, $T_s = 4000\,K$ is the surface temperature and $p^* = 8.5 \times 10^5\,Pa$ the vapor pressure. These values for chosen in such a way, that they correspond to the expected surface temperature pressure of the arc. The current density was assumed to be homogeneous over the arc root and to reproduce the current density of a realistic arc.

To validate the obtained Lagrangian finite element approach, the results are compared with those obtain using ANSYS Fluent. The main characteristics of the two solvers are highlighted in Table 22.1. The same initial mesh (with 21,898 nodes) has been used for both solvers.

Figures 22.2, 22.3, 22.4, 22.5, and 22.6 show velocities, density, pressure and temperature at time $t = 2\,ms$.

The results of the present approach can be seen on the left, and those obtained with ANSYS fluent are shown on the right. Although two different numerical approaches have been used, the results are very close and confirm the potential of the Lagrangian method proposed.

It is important to recall that the two solvers are based on very different techniques; in particular the Fluent solver implements a Finite Volume formulation with an implicit time integration on an Eulerian fixed mesh. The codes also have been run on different machines which makes the comparison in terms of CPU time extremely

Table 22.1 Comparison of the two solvers

	Present approach	Fluent
Equations	Lagrangian	Eulerian
Time integration	Explicit	Implicit
Space discretization	FEM (P1P1)	FV (2 order)
Mesh	Evolve in time	Fixed in time
Unknowns definition	Nodal based	Element based

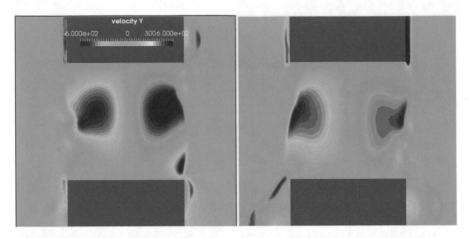

Fig. 22.2 Contour plot of the vertical velocity. On the left, the present approach; on the right, ANSYS Fluent. The same color scale is used

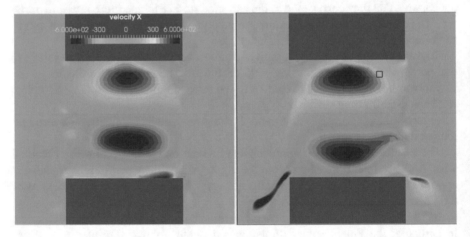

Fig. 22.3 Contour plot of the horizontal velocity. On the left, the present approach; on the right, ANSYS Fluent. The same color scale is used

difficult. The Lagrangian code (OpenMP parallelized runs on four Xeon cores) completed the simulation in roughly 10 h.

The qualitative and quantitative agreement is very good, taking into account the complexity of the overall model and of the formulation differences. The same color scales have been adopted to plot the results of the two codes. Even if the problem is defined on a symmetric domain and symmetric boundary conditions are used, symmetry in the solution is never enforced and some deviations develop. These are

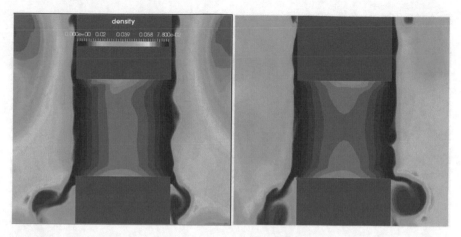

Fig. 22.4 Contour plot of the density. On the left, the present approach; on the right, ANSYS Fluent. The same color scale is used

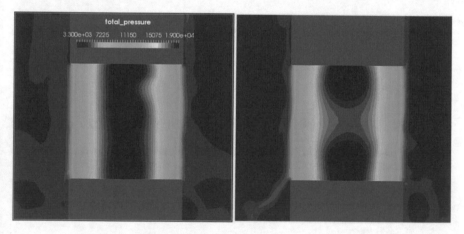

Fig. 22.5 Contour plot of the pressure. On the left, the present approach; on the right, ANSYS Fluent

possibly due to the unstructured and asymmetric mesh used, but also to intrinsic internal instabilities leading, e.g., to the density curls. It should also be noted that results are plotted at a given time, but actually the solution is time dependent due to these continuous symmetry breaking oscillations.

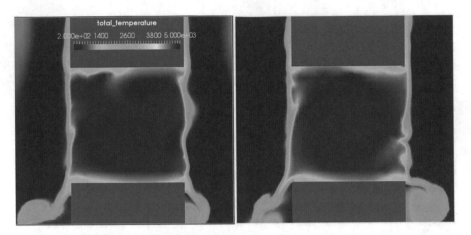

Fig. 22.6 Contour plot of the temperature. On the left, the present approach; on the right, ANSYS Fluent

Acknowledgements Financial support of ABB Corporate Research is gratefully acknowledged.

References

1. Slade, P.G.: The Vacuum Interrupter: Theory, Design, and Application. CRC Press, Boca Raton (2008)
2. Teichmann, J., Romheld, M., Hartmann, W.: Magnetically driven high current switching arcs in vacuum and low pressure gas. IEEE Trans. Plasma Sci. **27**(4), 1021 (1999)
3. Boxman, R.L.: High-current vacuum arc column motion on rail electrodes. J. Appl. Phys. **48**, 1885–1889 (1977)
4. Delachaux, T., Fritz, O., Gentsch, D., Schade, E., Shmelev, D.L.: Numerical simulation of a moving high-current vacuum arc driven by a transverse magnetic field. IEEE Trans. Plasma Sci. **35**, 905 (2007)
5. Delachaux, T., Fritz, O., Gentsch, D., Schade, E., Shmelev, D.L.: Simulation of a high current vacuum arc in a transverse magnetic field. IEEE Trans. Plasma Sci. **37**, 1386 (2009)
6. Shmelev, D.L., Delachaux, T.: Physical modeling and numerical simulation of constricted high-current vacuum arcs under the influence of a transverse magnetic field. IEEE Trans. Plasma Sci. **37**, 1379–1385 (2009)
7. Fritz, O., Shmelev, D., Hencken, K., Delachaux, T., Gentsch, D.: Results of 3D numerical simulations of high-current constricted vacuum arcs in a strong magnetic field. In: Proceedings of the 24th International Symposium on Discharges and Electrical Insulation in Vacuum (ISDEIV), 2010, pp. 359–364 (2010)
8. Cremonesi, M., Ferrara, L., Frangi, A., Perego, U.: Simulation of the flow of fresh cement suspensions by a Lagrangian finite element approach. J. Non-Newtonian Fluid Mech. **165**, 1555–1563 (2010)
9. Cremonesi, M., Ferrara, L., Frangi, A., Perego, U.: A Lagrangian finite element approach for the simulation of water-waves induced by landslides. Comput. Struct. **89**, 1086–1093 (2011)
10. Cremonesi, M., Frangi, A., Perego, U.: A Lagrangian finite element approach for the analysis of fluid–structure interaction problems. Int. J. Numer. Methods Eng. **84**, 610–630 (2010)

11. Scovazzi, G., Christon, M.A., Hughes, T.J.R., Shadid, J.N.: Stabilized shock hydrodynamics: I. A Lagrangian method. Comput. Methods Appl. Mech. Eng. **196**, 923–966 (2007)
12. Dobrev, V.A., Ellis, T.E., Kolev, T.V., Rieben, R.N.: Curvilinear finite elements for Lagrangian hydrodynamics. Int. J. Numer. Methods Fluids **65**, 1295–1310 (2011)
13. Spitzer, L.: Physics of Fully Ionized Gases. Interscience, New York (1965)
14. Cremonesi, M., Frangi, A.: A Lagrangian finite element method for 3D compressible flow applications. Comput. Methods Appl. Mech. Eng. **311**, 374–392 (2016)
15. Dobrev, V.A., Kolev, T.V., Rieben, R.N.: High-order curvilinear finite element methods for Lagrangian hydrodynamics. SIAM J. Sci. Comput. **34**, B606–B641 (2012)

Chapter 23
Virtual High Voltage Lab

**Andreas Blaszczyk, Jonas Ekeberg, Sergey Pancheshnyi,
and Magne Saxegaard**

Abstract A simulation framework for dielectric design of high voltage devices has been presented. It is based on a simplified modeling of discharge characteristics including propagation path, breakdown and inception voltages. A 3-stage procedure of evaluating surface charging as well as the numerical formulation of saturation charge boundary condition have been explained. The results of surface charging for a case study with an insulating barrier and the discharge evaluation for a complex gas insulated switchgear arrangement have been presented as application examples.

23.1 Introduction

During dielectric type tests, the voltage load specified in technical standards is applied to the tested device. The test is successful if either no breakdown of the insulation of the device occurs or, in the case of lightning impulse applied to gas insulated switchgear, only a limited number of breakdowns are observed. The test results are typically predicted by performing electrostatic field computations and comparing the calculated electric field strengths with the critical values specified for the relevant gases and surfaces of the geometry. This approach is very helpful, but in many cases we observe significant deviation from results obtained in the real high voltage lab.

In this paper we present a concept of a Virtual High Voltage Lab (VHVLab), which is aimed at closing the gap between simulations and experimental results.

A. Blaszczyk (✉) · J. Ekeberg
ABB Corporate Research, Baden-Dättwil, Switzerland
e-mail: andreas.blaszczyk@ch.abb.com; Jonas.Ekeberg@ch.abb.com

S. Pancheshnyi
ABB Transformers, Zürich, Switzerland
e-mail: Sergey.Pancheshnyi@ch.abb.com

M. Saxegaard
ABB Distribution, Skien, Norway
e-mail: Magne.Saxegaard@no.abb.com

© Springer International Publishing AG, part of Springer Nature 2018
U. Langer et al. (eds.), *Scientific Computing in Electrical Engineering*,
Mathematics in Industry 28, https://doi.org/10.1007/978-3-319-75538-0_23

The foreseen benefits are not only related to the accuracy of the predicted withstand voltages, but also to the fact that more insight into the discharge phenomena occurring during tests is expected from simulations. For example, the knowledge of a critical discharge path or surface charging in a complex geometry is of high interest to engineers.

VHVLab does not attempt to perform first principle simulations with their whole physical and numerical complexity. Instead we propose a simplified approach that evaluates different discharge stages including inception, streamer propagation, surface charging, leader transition and breakdown [1], while showing, in each of these stages, the corresponding discharge characteristics. The components of this approach as well as the overview of VHVLab concept is presented in Sects. 23.2 and 23.3. In Sects. 23.4 and 23.5 we show the utilization of VHVLab to predict dielectric tests of a real application with respect to inception as well as to explain charging phenomena occurring in a simple barrier arrangement.

23.2 Basic Concept

The architecture of the software concept is shown in Fig. 23.1. The core, consisting of discharge modeling procedures, is connected via predefined interfaces with external components, including the background field solver and the visualizer. As a first step of the analysis, engineers need to create the virtual CAD model of the device and compute it with an electrostatic solver that implements the VHVLab interface. Within the VHVLab session, the numerical computation of discharge characteristics in different gases, such as the streamer path or surface charge layer, is performed and presented to the user in a 3D visualizer.

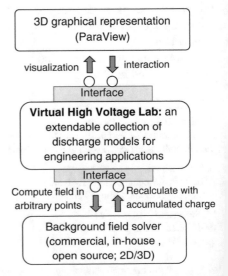

Fig. 23.1 Architecture of Virtual High Voltage Lab. **Comment:** An important feature of this architecture is the plug-in concept. VHVLab is not implemented as an add-on to any existing software system. It is a self-standing application that uses other commercial, open source and in-house components as plug-ins connected via interfaces

23.3   VHVLab Components

23.3.1   Gas Database

The main focus of VHVLab is gas insulation. Therefore, the electric discharge characteristics for different gases used as insulating medium in high voltage devices are included in VHVLab and can be selected by designers for analysis. The most important characteristics are the critical field E_{crit}, specifying the lowest field strength value for which the ionization process may start, as well as the effective ionization coefficient α_{eff}, which is a function of electric field strength and depends on gas type, pressure and temperature. Naturally, $\alpha_{eff}(E_{crit}) = 0$.

Experimentally measured α_{eff} is typically only available for simple gases (the online LXCat project [2] offers probably the most comprehensive collection of electron swarm data) and theoretical extension of these data to more complex cases, e.g. mixture of gases, often remains the only feasible approach. Among different possible techniques, solving the Boltzmann equation for the electron distribution in the so-called "two-term approximation" remains the most developed and used approach nowadays [3]. The complete sets of cross sections needed as input to a Boltzmann equation solver is also available in LXCat [2] for many atomic and molecular species.

It is generally accepted that inception of a gas discharge starts with the appearance of a first free electron in the gap. Depending on the conditions, such electrons can be emitted from a surface or can appear by detachment from a negative ion. A Townsend avalanche usually develops from the initiating electron. It is characterized by low space charge and, consequently, an insignificant electric field disturbance. The ionization coefficient α_{eff} allows to estimate the number of electrons created along the avalanche path, which consequently determines whether the self-sustaining discharge can be initiated. This can be expressed as a streamer criterion in the following form:

$$\int_x \alpha_{eff}(E)dx = \ln(N_e) > K_{str} \qquad (23.1)$$

where the path x starts at critical spots on a surface and follows the direction of the avalanche; N_e is the number of electrons created by the avalanche. If the logarithm of this number is larger than the streamer constant K_{str}, inception may occur. The exact value of K_{str} depends on the mechanism of discharge initiation [4–6]: the two most relevant cases are streamer corona, $K_{str} \simeq 18$ and glow corona, $K_{str} \simeq 10$.

With the introduction of fluorinated ketones and its mixtures with air, the so-called ABB AirPlus solution for eco-efficient medium voltage products [7], came the need to perform swarm measurements of the effective ionization coefficient α_{eff} and to assess the value of the inception streamer constant K_{str} using reference experiments.

23.3.2   Critical Spot Evaluation

A critical spot is defined as a region on the boundary between solids and gas domain in which all points have a field magnitude larger than E_{crit}. Evaluation of critical spots is based on mesh topology and results obtained from electrostatic background field computations loaded into VHVLab. The number of points (= mesh nodes) per spot can vary from a few up to many hundreds or thousands. In VHVLab it is possible to analyze all of them. However, in order to avoid excessive computation times, the user has the freedom to select only the points having the highest field magnitudes. These point are used for the analysis of discharge initiation as described in the next subsection.

23.3.3   Evaluation of Discharge Path and Its Characteristics

1. **Inception:** For a selected critical surface spot, we need the avalanche path in order to evaluate inception according to (23.1). Typically, this path can be calculated as a field line for which only the first few millimeters decide whether the inception occurs or not. The value of inception voltage U_{inc} is calculated iteratively by changing the applied voltage and scaling the E-field values along the path until the equality in relation (23.1) is achieved. In case of electrode-less inception (on dielectric surfaces), the magnitude of the tangential component can be used as a base to evaluate (23.1). Inception is a very important step in the prediction of high voltage tests. For weakly inhomogeneous fields, inception is the deciding mechanism of whether breakdown occurs or not so that the evaluation of further discharge phenomena can be neglected (except if surface charging is possible, since it can mitigate or enhance the inception).
2. **Propagation:** Once the streamer head has been created, we evaluate if the opposite electrode can be reached or if the discharge will be extinguished due to insufficient voltage. In particular for air, the stability criterion must be checked [5]. For example a positive streamer in air requires 0.5 kV/mm for stable propagation. Therefore, the proper estimation of the streamer length is crucial. A reliable engineering approach for evaluation of discharge path length is currently based on clearances between electrodes. A discharge-dependent path evaluation [1] still needs investigation before being applied in design.
3. **Leader transition and breakdown:** If the propagating streamer hits the dielectric surface, which is close to the electrode (like a coating), a leader transition may occur due to strong capacitive currents supporting the thermo-ionization of the creeping discharge. After leader transition, the discharge becomes unstable, which is equivalent to breakdown. The numerical prediction of leader characteristics is still a subject of research projects. Currently, engineering predictions are based on the leader inception voltage [5] specifying the highest voltage that can be applied across a dielectric layer without the risk of leader transition.

23.3.4 A 3-Stage Surface Charge Evaluation Procedure

The development of a discharge may influence the background field by generating space and surface charges. In particular, accumulation of charge on dielectric surfaces leads to significant changes of the discharge inception and propagation conditions. Therefore, a re-calculation of the background field, under presence of surface charge, may be required. In VHVLab, we proposed a computation of the maximum possible surface charge, called "saturation charge", as the next evaluation stage. Since it is a relatively new component, we describe the corresponding formulation together with an application example in Sect. 23.5. The computed saturation charge may remain on dielectric surfaces while the new load conditions appear (e.g. changed polarity). Therefore, the whole procedure may require more iterations. The VHVLab specifies three basic stages of background field computations, each of them followed by an evaluation of critical spots and discharge characteristics:

- **Stage 1:** Initial background field computation,
- **Stage 2:** Field computation with the same electric potential load as in Stage 1 but with the saturation charge boundary condition on dielectric surfaces that have been affected by the streamer propagation process evaluated within Stage 1,
- **Stage 3:** Field computation with the surface charge obtained during Stage 2 but with changed potential loads e.g. HV-electrode grounded (LI) or opposite polarity of HV-electrode (AC).

23.4 Application Example: Ring Main Unit

Ring Main Unit (RMU) is a switchgear component used for distribution of electric energy in the medium voltage power grid [8]. In order to achieve a compact size of the switching panels, the RMU has been traditionally designed using SF_6 gas for insulation. The latest trends to replace SF_6 by environmentally friendly but dielectrically less efficient gases have forced engineers to re-design the internal configuration of electrodes. An example is the newest design of the air-insulated 12 kV RMU panel where SF_6 has been substituted by air while keeping the same outer dimensions. For the optimization of electrode shapes and the final prediction of dielectric tests the engineers used the VHVlab approach as demonstrated in Fig. 23.2.

The test and simulation results are presented in Table 23.1. We consider here the original 12 kV air-insulated design as well as the same device but filled with AirPlus, which is supposed to enable operation at the voltage level of 24 kV. The test results represented by the withstand voltage $U_{2\%}$ is approximately 10% higher than U_{inc} predicted by simulation. Besides the statistical reasons, this deviation can be explained by the expected difference between the inception and the final breakdown voltage levels. However, designers accept this difference as a safety margin, which ensures that the increased test duty ($1.05 * U_{LI}$) can be passed. Consequently, we

Fig. 23.2 Simulation and test results for a disconnector of the Ring Main Unit (RMU): **(left)** Simulation model: snapshot of field distribution with critical spots and selected field lines in VHVlab visualizer **(right)** Tested object: traces of breakdowns on the surface of shielding electrodes after impulse test (denoted by circular lines). **Comments:** The critical spots are denoted here by black numbers. Each of these spots includes mesh nodes with $E > E_{crit}$ shown as white dots. For relevant spots the user selects in VHVLab the nodes with the highest field strength and calculates field lines (shown here as white trajectories). The initial part of a field line (up to a few millimeters) is used as a trajectory of avalanches when evaluating U_{inc} according to (23.1). The inception locations match well with the breakdown traces observed after experiments

Table 23.1 Simulation and test results for the Ring Main Unit (RMU)

U_{rated} kV	U_{LI}[a] kV	Gas @1.3 bar	E_{crit} kV/mm	Raw data of LI-test[b] Passed test	Failed test	$U_{50\%}$[c] kV	$U_{2\%}$[c] kV	U_{inc} kV
12	75	Air	3.3	84 kV/15/0	88 kV/11/4	88.6	86.7	77.8
24	125	AirPlus[d]	7.1	136 kV/14/1	141 kV/10/5	143.0	133.7	120

[a]U_{LI} is the lightning impulse voltage level required for type tests. Typically the device is tested for increased duty, which requires 5% higher U_{LI}
[b]In order to pass the LI test it is required that in a sequence of 15 impulses (shots) no more than 2 breakdowns occur. The test sequences have been repeated by increasing U_{LI} by 5% until the test failed. The raw data of the passed and the failed tests include the following information: applied impulse voltage/number of successful shots/number of breakdowns
[c]The breakdown $U_{50\%}$ and the withstand $U_{2\%}$ voltages have been evaluated based on the raw data of the both LI-tests (30 shots) using the maximum likelihood method and assuming normal distribution [9]
[d]A mix of air with 10.4% mole fraction of fluorinated ketones

conclude that for this type of weakly inhomogeneous fields, without the influence of surface charging, the U_{inc} can be reliably predicted with VHVLab. Applying the design criterion $U_{inc} >= U_{LI}$ enables successful tests.

23.5   Surface Charging

23.5.1   Saturation Charge Formulation

Saturation is an extremal stage of charge accumulation on a dielectric surface (Stage 2). We assume that the amount of accumulated charge does not allow streamers, propagating onto or along dielectrics, to deposit more charge on the saturated surface. For the interface between a solid dielectric and gas, we formulate two equations describing saturation and continuity boundary conditions as follows:

$$E_{nGas} = 0 \tag{23.2}$$

$$E_{nDiel}\epsilon_{Diel} = \sigma_{sat} \tag{23.3}$$

where E_{nGas} and E_{nDiel} are normal components of the field strength at the charged surface in the gas and inside the barrier respectively, ϵ_{Diel} is the permittivity of the solid dielectric, whereas σ_{sat} is the unknown saturation charge density accumulated on the surface. The initial background field computation (Stage 1) as well as the computation with known surface charge (Stage 3) uses only the continuity boundary condition in the following form:

$$E_{nDiel}\epsilon_{Diel} = E_{nGas}\epsilon_{Gas} + \sigma_s \tag{23.4}$$

where σ_s is the known charge accumulated on the dielectric surface: for Stage 3 $\sigma_s = \sigma_{sat}$ (σ_{sat} calculated in Stage 2), whereas for the initial Stage 1 $\sigma_s = 0$.

The formulation defined by (23.2) and (23.3) is a non-standard feature that needs to be implemented for electrostatic solvers integrated within VHVLab. This implementation is straightforward for the indirect integral formulations of electrostatic problems [10, 11]. In spite of increased dimension of the equation system due to additional unknowns (σ_{sat}), the computational effort is still moderate and fully applicable in the engineering environment.

23.5.2   Case Study: Rod-Barrier Arrangement

We consider a simple arrangement including a 1.6-mm thick dielectric barrier ($\epsilon_r = 3.0$) placed between a high voltage rod (7-mm diameter) and a grounded plane [12], see axisymmetric representation for Stage 1 in Fig. 23.3a. Discharges are created at the rod tip during the positive lightning impulse of $U_{LI} = 91\,kV$ since inception voltage is only $U_{inc} = 33\,kV$. These discharges propagate towards the barrier and deposit positive charge on the upper surface of the barrier. Assuming that enough charge has been delivered from the rod, a state of saturation may arise. The surface charge mitigates the high field strength at the tip of the rod but discharges will

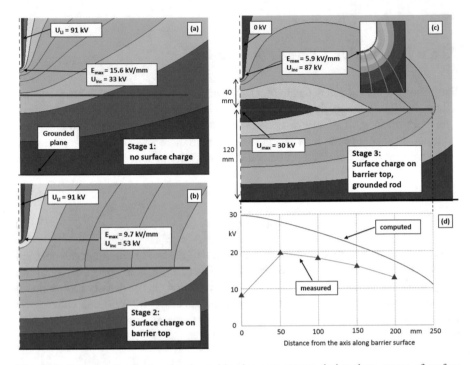

Fig. 23.3 Potential distribution for the rod-barrier arrangement during three stages of surface charging: (**a**) without surface charge on the barrier and $U_{LI} = 91$ kV at the rod; (**b**) with saturation charge on the barrier and the unchanged LI-voltage at the rod; (**c**) with saturation charge and zero-potential at the rod; (**d**) comparison of the voltage distribution along the barrier surface with the measured values

occur since the inception voltage of 53 kV is still lower than the applied 91 kV, see Fig. 23.3b (in this case the saturation charge does not prevent the inception). The withstand voltage for stages 1 and 2 can be reliably predicted by applying the streamer stability criterion for the discharge path determined by the clearance between the rod tip and the grounded plane [12, 13].

Once the saturation charge has been obtained, it can be applied as a load during the next stage when the potential of the rod is back to zero. The corresponding voltage distribution for this stage is shown in Fig. 23.3c. The measured and the calculated barrier potential differs significantly within the area just below the zero-potential rod, Fig. 23.3d. Besides the influence of space charge and ion drifting, this difference can be explained by the fact that the computation does not include the re-strike phenomenon that may occur due to the high stress (5.9 kV/mm) at the rod tip. Analysis of inception at the rod showed that the saturation charge would be sufficient to trigger a re-strike in air for impulse voltages above 87 kV. This value is lower than the applied 91 kV and negative streamers may therefore start at the grounded tip and recombine with the positive charge accumulated on the barrier.

23.6 Conclusion

Virtual High Voltage Lab provides a comprehensive simulation framework that is focused on simulation-based prediction of high voltage tests required in the development of power devices. It integrates physical principles, advanced numerical methods and empirical engineering knowledge in a tool that can be effectively used in the industrial engineering environment. The goal of VHVLab is to replace prototype testing by virtual tests and consequently to reduce development costs, as well as to significantly improve product quality thanks to a better understanding of discharge phenomena.

References

1. Pedersen, A., Christen, T., Blaszczyk, A., Böhme, H.: Streamer inception and propagation models for designing air insulated power devices. In: IEEE CEIDP Conference Material, Virginia Beach, October 2009
2. LXCAT Plasma Data Exchange Project. www.lxcat.net
3. Hagelaar, G.J.M., Pitchford, L.C.: Solving the Boltzmann equation to obtain electron transport coefficients and rate coefficients for fluid models. Plasma Sources Sci. Technol. **14**, 722 (2005)
4. Petcharaks, K.: Applicability of the streamer breakdown criterion to inhomogenous gas gaps. Ph.D. Thesis No. 11192. ETH Zurich, 1995
5. Kuechler, A.: Hochspannungstechnik: Grundlagen – Technologie – Andwendungen. VDI-Verlag, Düsseldorf (1996). ISBN: 3-18-401530-0
6. Kuffel, E., Zaengl, W.S., Kuffel, J.: High Voltage Engineering: Fundamentals. Elsevier, Amsterdam (2000)
7. Hyrenbach, M., Hintzen, T., Müller, P., Owens, J.: Alternative gas insulation in medium-voltage switchgear. In: 23rd International Conference on Electricity Distribution, Lyon, 15–18 June 2015
8. Saxegaard, M., Kristoffersen, M., Stoller, P., Seeger, M., Hyrenbach, M., Landsverk, H.: Dielectric properties of gases suitable for secondary medium voltage switchgear. In: 23rd International Conference on Electricity Distribution, (CIRED), Lyon June 2015, paper 0926
9. Hauschild, W., Mosch, W.: Statistical Techniques for High-Voltage Engineering. Power Series, vol. 13. The Institution of Engineering and Technology, London (2007)
10. Blaszczyk, A., Steinbigler, H.: Region-oriented charge simulation. IEEE Trans. Magn. **30**(5), 2924–2927 (1994)
11. De Kock, N., Mendik, M., Andjelic, Z., Blaszczyk, A.: Application of 3D boundary element method in the design of EHV GIS components. IEEE Mag. Electr. Insul. **14**(3), 17–22 (1998)
12. Ekeberg, J.: Dielectric multi-barrier flashovers in air. ABB CH-RD 2015-61 Internal Report on Experimental Investigations (2015)
13. Mauseth, F., Jorstad, J.S., Pedersen, A.: Streamer inception and propagation for air insulated rod-plane gaps with barriers. In: IEEE CEIDP Conference Material (2012)

Index

circuit breaker, 50
co-simulation, 81
coupled problems
    electro-thermal, 44, 190

droplet dynamics, 101
dynamical system, 158, 190, 204

electric motor, 154
electromyography, 14
equations
    circuit, 114
    differential-algebraic, 81, 92, 114, 138
    Laplace, 7
    Maxwell, 27, 36, 91, 180
    Maxwell-Ampere, 26, 27
    Navier-Stokes, 102

finite-integration technique, 27

generalized Gaussian distributions, 170

interface problems
    mesh adaption, 148, 223

method
    adaptive finite element, 46
    cubature, 158
    discontinuous Galerkin, 128

explicit/forward Euler, 139
finite element, 7, 14, 138, 219
finite integration, 27
finite volume, 27
implicit/backward Euler, 57, 222
multirate shooting, 114
multirate time integration, 94
multistep, 114
Newton's, 57, 114
preconditioned conjugate gradient, 139
quasi Monte Carlo, 158, 204
shooting, 114
sparse grid quadrature, 204
Trefftz, 131
model
    approximate ladder-circuit, 236
    bio-electric, 5
    equivalent circuit, 180
    magnetodynamic levitation, 221
model order reduction, 93, 190, 209, 222

nodal analysis, 17

optimization
    robust, 183
    shape, 14, 147
    topology, 147

polynomial chaos expansion, 56, 60
problem
    eddy current, 47, 127, 138, 219
    electro-quasistatic, 7

    magnetoquasistatic, 91, 138
    multiphase flow, 102
    RFIC isolation, 178
    TEAM 10, 141
    TEAM 28, 220, 233
proper orthogonal decomposition, 222

sensitivity analysis, 36, 56, 157
simulation
    coupled circuit device, 70

        electromagnetic transient, 26
        multirate DAE/ODE, 91
skin effect, 128

uncertainty quantification, 55, 182

vacuum arc, 244
vacuum interrupter, 243
Virtual High Voltage Lab, 256

Printed in the United States
By Bookmasters